Aquaculture

NIPA® GENX ELECTRONIC RESOURCES & SOLUTIONS P. LTD.
New Delhi-110 034

About the Editors

Durgesh Kumar Verma completed B.F.Sc. & M.F.Sc. in Fisheries Resource Management from the Acharya Narendra Deva University of Agriculture and Technology, Kumarganj Ayodhya, India. He has authored/published several research and review papers, popular articles, chapters, and books. He is an active and dedicated member of several professional societies mainly engaged in fisheries research and development in the country. He receives several accolades and awards such as Junior Researcher Award, Young Scientist Award, Best Paper Presentation awards, etc. at national and International meetings. Presently he is working as a young researcher at ICAR-CIFRI, Prayagraj, Uttar Pradesh.

Dr. Sanjay Kumar Gupta, M.F.Sc., Ph.D., has published many research and review papers in both national and international peer-reviewed journals. He has written/edited 15 books and 35 book chapters. He visited Australia, Belgium, Holland, France, Thailand, Indonesia and Malaysia. Dr. Gupta has presented papers in diverse national and international symposia, workshops, seminars and conferences. He is the recipient of a prestigious endeavour postdoc fellow by the Australian Government (2018). He has been honored with several academic and professional awards for his outstanding contribution in the field of aquaculture and fisheries. Presently, he is working as a Senior Scientist at ICAR-Indian Institute of Agricultural Biotechnology, Ranchi, Jharkhand, India.

Dr. Ashfauq Farooq Aga is a distinguished scientist with B.FSc, M.FSc. & Ph.D. specializing in aquaculture, affiliated with the Division of Fisheries at the Faculty of Fisheries SKUAST-K. He has made significant contributions to the field through his research and academic endeavours'. Aga has served as the editor of several notable books in the domain of fisheries, showcasing his expertise and leadership in compiling and disseminating knowledge in this specialized area of study. He is recipient of best scientist award by PFGF Jammu & Kashmir. He has participated and presented papers in diverse conferences, workshops, and seminars related to aquaculture R&D and education. His biodata reflects his academic achievements, research interests, and leadership in the field of fisheries science.

Dr. Sudeshna Sarker, HOD, Aquatic Animal Health Management, CoF, Kishanganj, BASU is an experienced researcher with a demonstrated history of working in the field of Aquatic Animal Health. Dr. Sarker secured "Dr. N.R. MENON Award" for the 'Best Indian M.F.Sc Thesis-2016' by PFGF, India, and is recipient of International Travel Support (ITS) by SERB, GOI for participating in the 5th International conference on "FLAVOBACTERIUM 2018, Japan. She has published research papers, several book chapters, popular articles, and edited a book. She has earlier worked as an Assistant Professor in the School of Agriculture and Allied Sciences, The Neotia University. Currently, Dr. Sarker is posted as Assistant Professor-cum-Junior Scientist in the Dept. of Aquatic Animal Health Management, under Bihar Animal Sciences University.

Aquaculture
Technological Advancements

Durgesh Kumar Verma
Young Researcher
ICAR-CIFRI, Prayagraj
Uttar Pradesh, India

Sanjay Kumar Gupta
Senior Scientist
ICAR-Indian Institute of Agricultural Biotechnology
Ranchi, Jharkhand, India

Ashfauq Farooq Aga
Assistant Professor
Division of Fisheries
Sher-e-Kashmir University of Agricultural Sciences and Technology
Shalimar, Srinagar, Jammu and Kashimir, India

Sudeshna Sarker
Assistant Professor and Head
Department of Aquatic Animal Health Management
College of Fisheries, Bihar Animal Sciences University
Bihar, India

NIPA® GENX ELECTRONIC RESOURCES & SOLUTIONS P. LTD.
New Delhi-110 034

NIPA® GENX ELECTRONIC
RESOURCES & SOLUTIONS P. LTD.

101,103, Vikas Surya Plaza, CU Block
L.S.C. Market, Pitam Pura, New Delhi-110 034
Ph : +91-11-43860225, Mob.: +91 9717133558, 9540816132
E-mail: newindiapublishingagency@gmail.com
Website: www.nipaersources.com

Print ISBN: 978-93-58878-69-1
ebook ISBN: 978-93-58879-31-5

Composed and Designed by NIPA®.

Preface

Welcome to "**Aquaculture: Technological Advancements**", a comprehensive exploration of the cutting-edge developments that are reshaping the future of fisheries and aquaculture. As the global demand for aquatic products continues to soar, therfore the need for innovative and sustainable approaches to ensure the health and productivity of aquatic systems is increasing exponentially. This volume serves as a beacon for researchers, practitioners, and enthusiasts seeking to understand the latest technological strides and their implications for the industry.

In this book, we delve into a range of topics that reflect the vibrant and dynamic field of aquaculture. Our journey begins with an examination of the recent advancements in biotechnology and molecular biology tools, which are revolutionizing the way we approach fisheries and aquaculture. These advancements are not only enhancing our understanding of aquatic species but also enabling more precise and effective management practices.

We then explore modern biotechnological approaches to fish-based biomaterials, highlighting the innovative uses of these materials in various applications. The blue biotech revolution, driven by next-generation sequencing technologies, is transforming our ability to sequence and analyze aquatic genomes, offering new opportunities for improving fish stocks and aquaculture practices.

The future of environmental DNA (eDNA) in aquaculture is also a focal point of discussion. While eDNA holds promise for monitoring aquatic environments and detecting species, small-scale farmers face unique challenges in adopting this technology. Our examination of these challenges aims to provide insights into how eDNA can be made more accessible and beneficial for all levels of aquaculture.

The bio-floc fish farming method, particularly its status in India, is another critical topic covered in this book. Bio-floc technology offers a sustainable approach to fish farming, and understanding its current application in different regions can provide valuable lessons and strategies for broader implementation.

We also address the application of medicinal plants in fish feed, discussing both its potential benefits and the challenges faced in integrating these plant extracts into aquaculture diets. Similarly, the application of recent tools in fish disease diagnosis and the limitations of fish therapeutics have been explored, emphasizing the need for continued research and development in these prominent areas.

The role of herbs in fish feed, the use of aquatic algae for controlling pesticide pollution, and the recent trends in fish farming are all discussed, providing a holistic view of the advancements and challenges in the field. Zero tillage fish farming and its importance, as well as the therapeutic impact of marine organisms on osteoporosis using machine learning, are also examined, highlighting the diverse ways in which technology is advancing aquaculture.

Finally, we delve into the carbon and nitrogen ratios in shrimp aquaculture systems and the challenges associated with recirculatory aquaculture systems (RAS). Histological alterations in fish as diagnostic tools for ecotoxicological research are also explored, thus underscoring the critical role of scientific innovation in monitoring and improving aquatic health.

This book is intended to serve as both a reference and a source of inspiration for those engaged in the study and practice of modern aquaculture. By bringing together leading experts from multiple organisations and their latest research compilation, we hope to foster a deeper understanding of the technological advancements shaping the future of the industry.

We invite you to explore these topics and join us in the journey towards a more sustainable and technologically advanced aquaculture sector.

Sincerely

Editors

Contents

List of Colour Plates

Abbreviations

3d	Three-Dimensional
ABV	Aquabirnavirus
ACV	Apple Cider Vinegar
AFLP	Amplified Fragment Length Polymorphism
AFP	Antifreeze Protein
AI	Artificial Intelligence
AICRP	All India Coordinated Research Project
AMR	Antibiotic Resistance
APX	Ascorbate Peroxidase
ARMS-PCR	Amplification Refractory Mutation System-Polymerase Chain Reaction
ASE	Accelerated Solvent Extraction
AUROC	Area Under The Receiver Operating Curve
BCF	Bioconcentration Factor
BGA	Blue-Green Algae
BKC	Benzalkonium Chloride
BKD	Bacterial Kidney Disease
BRONJ	Bisphosphonate-Related Osteonecrosis of The Jaw
BTF	Biofloc Technology
BTS	Benchtop Sequencers
C/N	Carbon/Nitrogen
CAGR	Compound Annual Growth Rate
CAT	Catalase
cDNA	Complementary DNA
ChiP-Seq	Chip Sequencing
CIFRI	Central Inland Fisheries Research Institute
CMFRI	Central Marine Fisheries Research Institute
CNP	Cinnamon Nanoparticles
CO_2	Carbon dioxide
Cr	Chromium
Crispr	Clustered regularly interspaced short palindromic repeats
CTAB	Hexadecylcetyl Trimethyl Ammonium Bromide
CuSO4	Copper Sulfate
DFA	Direct Fluorescent Antibody
DNA	Deoxyribonucleic Acid
DO	Dissolved Oxygen
ECM	Extracellular Matrix
eDNA	Environmental DNA
EEZ	Exclusive Economic Zone
ELISA	Enzyme-Linked Immunosorbent Assay
ESC	Embryonic Stem Cell

FAO	Food and Agriculture Organization
FAT	Fluorescence Antibody Techniques
FB	Emamectin Benzoate
FCR	Feed Conversion Rate
FCR	Feed Conversion Ratio
FDA	Food and Drug Administration
FISH	Fluorescence In Situ Hybridization
FSH	Follicular Stimulating Hormone
GDP	Gross Domestic Product
GE	Genome Editing
GELMA	Gelatin Methacrylate
Gly	Glycine
GMOs	Genetically Modified Organisms
GMPS	Good Management Practices
GnRH	Gonadotropin Releasing Hormone
GS	Genomic Selection
$H_2 O_2$	Hydrogen Peroxide
HA	Hyaluronic Acid
HFIP	Hexafluoro-Isopropanol
HMGBs	High Mobility Group Box Proteins
HPV	Hepatopancreatic Parvovirus
HSP	Heat Shock Proteins
HWE	Hot Water Extraction
Hyp	Hydroxyproline
ICAR	Indian Council of Agricultural Research
IFA	Indirect Fluorescent Antibody
IFN-γ	Interferon-Gamma
IHNV	Infectious Hematopoietic Necrosis Virus
IL-1β	Interleukin 1βeta
IL-8	Interleukin 8
IMTA	Integrated Multi-Trophic Aquaculture
Indels	Insertions and deletions
IPNV	Infectious Pancreatic Necrosis Virus
ISH	In Situ Hybridization
ISKNV	Infectious Spleen and Kidney Necrosis Virus
$KMnO_4$	Potassium Permanganate
LCA	Life-Cycle Analysis
LDV	Lymphocystis Disease Virus
LH	Luteinizing Hormone
MAS	Marker-Assisted Selection
miRNA	MicroRNA
MMT	Million Metric Tons
MPSS	Massively Parallel Signature Sequencing
mRNA	Messenger RNA
MRSA	Methicillin-Resistant Staphylococcus aureus
Ms	Metagenomic Sequencing
MSSA	Methicillin-Sensitive Staphylococcus aureus
N	Nitrogen
NaCl	Sodium Chloride

NCHI	National Center For Biotechnology Information
NGO	Non-Governmental Organization
NGS	Next-Generation Sequencing
NH_3	Ammonia
O_2 -	Superoxide Anion
OCPs	Organochlorine Pesticides
OH	Hydroxyl Radical
ONJ	Osteonecrosis of The Jaw
ORs	Odds Ratios
P	Phosphorus
PCL	Polycaprolactone
PCR	Polymerase Chain Reaction
PMMSY	Pradhan Mantri Matsya Sampada Yojana
Pro	Proline
PRRs	Pattern Recognition Receptors
PSS	Production Scale Sequencers
PVC	Polyvinyl Chloride
QS	Quorum Sensing
QS	Quorum Signaling
QTL	Quantitative Trait Locus Mapping
QTLs	Quantitative Trait Loci
RAPD	Random Amplified Polymorphic DNA
RAS	Recirculating Aquaculture Systems
REase/ENase	Restriction Endonuclease
RF	Random Forest
RF	Random Forest
RFLP	Restriction Fragment Length Polymorphism
RIG-I	Retinoic Acid Inducible Gene-I (Rig-I)
RLRs	Retinoic Acid Inducible Gene-I–Like Receptors
RNA	Ribonucleic Acid
ROS	Reactive Oxygen Species
RT-PCR	Reverse Transcriptase PCR
SBS	Sequencing By Synthesis
SCF	Supercritical Fluid
SCPs	Single-Cell Proteins
SDG	Sustainable Development Goal
SGR	Specific Growth Rate
SM	Secondary Metabolites
SMs	Secondary Metabolites
SNPs	Single-Nucleotide Polymorphism
SNV	Single Nucleotide Variations
SOD	Superoxide Dismutase
SOLiD	Sequencing by Oligonucleotide Ligation and Detection
sRNA	Small RNA
SV	Structural Vaccinology
SVM	Support Vector Machine
TA	Total Alkalinity
TAN	Total Ammonia Nitrogen
TGEP	Target Gene Expression Profiling

TGF-β	Transforming Growth Factor-Beta
TLRs	Toll-Like Receptors
TMS	Tricaine Methane Sulfonate
TNF- α	Tumor Necrosis Factor-Alpha
UV	Ultraviolate
VEGF-A	Vascular Endothelial Growth Factor A
VFD	Veterinary Feed Directive
VHSV	Viral Haemorrhagic Septicaemia Virus
WG	Weight Gain
WGS	Whole Genome Sequencing
ZMWs	Zero-Mode Waveguides

1

Recent Advances of Biotechnology and Molecular Biology Tools in Fisheries and Aquaculture

Aishwarya Sharma[1*], Aditya Pratap Acharya[2], Sagarika Swain[1] and Sarvendra Kumar[1]

[1]*College of Fisheries, Kishanganj, Bihar Animal Sciences University, Patna - 855107 Bihar, India*

[2]*Faculty of Veterinary & Animal Sciences, Mohanpur Campus, West Bengal University of Animal and Fishery Sciences, Kolkata - 700037, West Bengal, India*

Abstract

There is an ever-accelerating decline in the world's wild fish populations. The sole approach to guarantee the world with sufficient fish species is through innovations in the field of aquaculture. The history of conventional aquaculture dates back 4,000 years to China, where it has flourished during the last 30 years. Notwithstanding its potential benefits, aquaculture has encountered significant obstacles such as restricted improved species, labor-intensiveness, and contamination of the environment, diseases, and untraceability of goods. Novel technologies must be developed in aquaculture to boost the fish production. Innovative and revolutionary technologies such as genome editing, artificial intelligence, alternative proteins and oils to replace fish meals and fish oils, oral vaccination, might present possibilities for profitable and sustainable aquaculture. Additionally, aquaculture contributes significantly to the economies of rural areas by generating additional employment opportunities. The aquaculture sector is presently tasked with finding solutions to the concurrent issues of creating commercially feasible production methods, lessening environmental effect, and enhancing the public's opinion. Aquaculture has the potential to fulfill rising demands, and biotechnology can significantly increase aquaculture production. The field of biotechnology has several promising applications that have the potential to enhance human welfare via aquaculture through selective breeding, hybridization, health management, nutrition, cryopreservation, and conservation of fish genetic resources. This study provides a quick overview of these technologies in order to provide a space for a thorough conversation on how to incorporate them into aquaculture to increase its profitability and sustainability.

Keywords: *Selective breeding, Aquaculture, Nanotechnology, Reverse vaccinology, Surrogacy*

1. Introduction

Aquaculture has played a sustaining role in global food systems by providing expanded food production and livelihood benefits with relatively minimal environmental harm. (Igwegbe *et al.,* 2021). To emphasize the importance of the aquaculture sector, the Food and Agriculture Organization of the United Nations (2020) stated that, global aquaculture production increased by 5.27%, hitting a record-breaking productivity value of 114.5 million tonnes between 1990 and 2018 (Pauly and Zeller 2017). Aquaculture has profited from developments in science and technology in nearly every aspect such as a greater diversity of species, nutritional regimes, production techniques, organizational frameworks, and marketing compared to other agricultural industries (FAO 2020). As an illustration, enhanced reproductive technologies have made it possible for individuals to achieve closed life cycles of aquaculture species, resulting in the diversification of species in the aquaculture sector (Weber and Lee 2014). The utilization of live feeds in hatcheries, such as microalgae, artemia, rotifers, brine shrimp, and various copepods, has eliminated the obstacle in breeding a variety of marine organisms (Conceição *et al.,* 2010).

Approximately 60 varieties of fish species have shown significant improvements in features of economic value through selective breeding assisted by quantitative genetics (Zuma 2022). River shrimps (Levy *et al.*, 2017), Yellow catfish (Roskoski *et al.*, 1975), and mono-sex tilapias (Mair *et al.*, 1997) all have been produced owing to sex reversal technology and DNA markers linked to sex determination. The risk of inbreeding has decreased due to intrafamily screening being made possible by molecular parentage assignment (Xu *et al.*, 2020). Selection for characteristics, particularly are governed by single genes and a small number of significant genes (Fuji *et al.*, 2007; Houston *et al.*, 2008) has been made possible by QTL (quantitative trait locus mapping) and marker-assisted selection (MAS) (Yue 2014). Enhanced feed formulations dependent on the dietary needs for every fish species have enhanced feed conversion rate (FCR) and decreased feed cost (Tacon and Metian 2015). Innovations for disease control (Kelly and Renukdas 2020) have minimized the prevalence of ailments in aquaculture. Despite the fact that aquaculture has expanded tremendously as a result of these and several other pioneering discoveries, the difficulties confronting the industry are immense in order to fulfill the growing demand for seafood to the world's expanding population (FAO 2020). Deteriorating state of the environment, diminishing availability of fish meals and oils, and global warming would substantially impair our potential to manufacture sufficient aquaculture commodities to fulfill the growing market for seafood (Abdelrahman, *et al.*, 2017; Shen, *et al.*, 2021). Fish health analysis and stabilization might benefit greatly from the application of polymerase chain reaction (PCR). PCR-based diagnostic techniques and gene probes have been invented to diagnose a number of infections which impact shrimp and fish (Karunasagar and Karunasagar 1999). In addition to exerting an immediate favorable influence on a variety of the key components of fish health management, biotechnology also has an indirect impact on a number of other significant challenges. In order to efficiently control disease and minimize the use of pesticides and antibiotics in the environment, it is essential to identify pathogens promptly. Vaccination additionally assists in preventing disease while employing fewer drugs (Hill 2005).

In contrast to diagnostic probes and tests, substantial advancement has been achieved in the last ten years in the production of fish vaccines (Thompson and Adams 2004; Cunningham 2004). There are currently several cutting-edge technologies accessible to support the aquaculture sector and enhance fish health. Despite the fact that these treatments increased production, they had a negative impact on the environment and became hazardous. Sustainable development has benefited from new discoveries in biotechnology that are aimed at preserving and safeguarding the environment. Both the excessive production of waste and the careless use of feed; when these substances are discharged into the environment unattended, they contaminate the water, induce eutrophication, transmit disease, and worsen environmental conditions. Other consequences encompass biodiversity shifts, salinization of nearby soils, depletion of wild populations, and loss and modification of natural ecosystems. In order to provide readers a comprehensive understanding of the possible solutions via novel technologies, this study briefly describes and explores these emerging technologies and their potential applications to enhance the aquaculture sector.

2. Biotechnology in Fish Breeding

With respect to the concept of induced breeding of fishes, gonadotropin releasing hormone (GnRH) is currently the most effective biotechnological innovations readily accessible. According to Bhattacharya, *et al.* (2002), GnRH is the primary modulator and core activator of the reproductive pathway in every vertebrate species (Bhattacharya *et al.*, 2002). Initially recognized in pig and sheep hypothalami, this decapeptide possesses the potential to stimulate the pituitary processing of follicular stimulating hormone (FSH) and luteinizing hormone (LH) (Carolsfeld, *et al.*, 2000). Hence, accordingly, it has been determined that the single neuropeptide responsible for the production of LH and FSH in the majority of placental mammals, encompassing humans, is GnRH in its single state. Nonetheless, twelve GnRH variations in non-mammal species (apart from guinea pigs) had previously been structurally characterized; with seven or eight distinct versions of these variants have been identified from fish species (Carolsfeld, *et al.*, 2000; Halder, *et al.*, 1991). The latest study on GnRH purification and characterization was carried out by Tang *et al.* (2023). Numerous chemical analogs have been created based on their structural variations and biological functioning. One such analogue is the salmon GnRH analogue, which is widely utilized in fish breeding nowadays and is marketed commercially underneath the brand name "Ovaprim" worldwide. With the advancement of GnRH technology, fish may now be effectively bred artificially. The introduction of GnRH technology has made it possible to effectively breed fish in captivity (Halder, *et al.*, 1991).

3. Fish Feed Biotechnology

Fish meal is still one of the most widely used sources of protein in many diets that include fish owing to its abundant protein contents and superior nutritional value. By improving food palatability and boosting nutrient uptake, and absorption, adding fish meal to a diet also improves feed efficiency, quality, and accelerates growth. However, the usage of fish meal comes with certain drawbacks, major being its high cost. The cost of fish meal has considerably increased, reaching over US $1600 per metric ton. Additionally, it is derived from the leftovers of wild fish; causing fish populations to drop globally. It has considerably higher concentrations of phosphorus compared to what fish demand for a normal growth process. Penetration of a surplus of phosphorus in water, can lead to eutrophication (Adelizi 1998).

In aquaculture, particularly in finfish mariculture, Salmon's feed primarily depend on fish meal and fish oil (Han, *et al.*, 2018). Different species that are captivated from oceans, such as herring, krill, along with other minor forage species, are capable of producing byproducts like fish meal and fish oil. According to Hodar *et al.* (2020), fish meal possesses an exceptionally high protein concentration (Hodar *et al.*, 2020). The rapid growth of the aquaculture sector and the expanding market for farmed marine products have resulted in an increase in the price and quantity of fish meal and fish oils over the past few years (Cao *et al.*, 2015). Overfishing has also imposed considerable burden on natural populations of fishes. Fish meal resources are insufficient to adequately meet the aquaculture industry's requirements at the prevailing current pace of productivity (Hodar *et al.*, 2020). Substitute protein sources have been widely researched and explored as a potential replacement for fish meals. According to Hodar *et al.* (2020), there has been much research on plant-based proteins, such as soybean protein with beneficial outcomes (Hodar *et al.*, 2020). In substitute of fish meals, micro- and macroalgae have been incorporated to fish diets. According to Han *et al.* (2019), superior algae feed continues to be costly at the moment, but its effects are encouraging (Han, *et al.*, 2019). Fish meal can also be substituted with proteins derived from insects. Black soldier fly and crickets (Rumbos *et al.*, 2019) constitute the appealing possibilities for insect-based proteins (Mousavi, *et al.*, 2020). According to Wang and Shelomi (2017), there are reported techniques for cultivating these insects utilizing food wastes (Wang and Shelomi 2017). In an attempt to minimize expenses, a number of organizations have commenced manufacturing these insects and boosted their production. Jones *et al.* (2020) recognized single-cell proteins (SCPs) as an additional category of alternative proteins (Jones *et al.*, 2020). According to Sharif *et al.* (2020), bacteria, fungus, and algae develop SCPs (Sharif *et al.*, 2020) and these SCPs may be able to fulfil the demands for protein in the aquafeed sector. According to feeding research studies, SCPs may substitute for the place of fish meals in Atlantic salmon, Rainbow trout, and White-leg shrimp (Jones *et al.*, 2020). As such, the SCPs show great potential as a fishmeal substitute. Considerable progress has been achieved in the last several decades in substituting plant oils replacing fish oil in fish feed formulations (Nasopoulou and Zabetakis 2012). Potential substitutes for fish oil include plant oils like rapeseed and palm. There are a few important factors that need to be taken into account, even though substitute proteins and oils seem promising for fish meals. A few of the issues mentioned above includes price, output capability, and supply stability.

Probiotics are typically consumed by animals as live microbial feed supplements. They work by enhancing the intestinal microbial equilibrium and enhancing the diversity of non-toxic species prevalent in the animal's diet. Specifically through the gastrointestinal system, a stable gut microbiota promotes the host's defense against pathogenic incursions. Aquaculture has recently started off to employ probiotics, despite their widespread usage in animal farming. Since, opportunistic bacteria like the luminous Vibrio harveyi have obstructed shrimp farming, there are growing reports concerning potential probiotics, and in certain circumstances, probiotics have been shown to substantially decrease the application of antibiotics in shrimp hatcheries. It has been demonstrated that it is possible to impede the growth of some harmful bacteria (like Vibrio spp.) in shrimp hatcheries through the implementation or inoculating non-pathogenic strains or species of bacteria that are antagonistic for microbial metabolite availability. This process appears to be both cost-effective and efficient (Lauzon *et al.*, 2014; Lazado and Caipang 2014). Probiotic bacteria have the ability to monitor and regulate the release of pro- and anti-inflammatory cytokines that are significant chemical messengers that impact immune cell monitoring, stimulation, development, and differentiation. Lactobacillus plantarum subsp. plantarum, for example, was found to be efficient in providing resistance to Lactococcus garvieae infection in rainbow trout by stimulating the expression of multiple cytokines in the kidney and gut, as well as upregulating the expression of IL-8 in the intestines (Pérez-Sánchez *et al.*, 2011). In the intestines of Olive flounder (Paralichthys olivaceus), dietary supplementation of Lactococcus lactis and Lb. plantarum additionally demonstrated an upsurge of cytokine gene expression and an enhanced susceptibility to Streptococcus iniae infection (Beck *et al.*, 2020). Comparable outcomes have also been shown in crustaceans, where dietary administering Bacillus subtilis strains, enhanced disease resistance of white shrimp, Litopenaeus vannamei and levied the expression of genes linked to the immune system (Zokaeifar *et al.*, 2020).

4. Biotechnological innovations for genetic improvement

Aquaculture's global expansion has been largely attributed to genetic enhancement via breeding. According to Gjedrem and Robinson (2014), conventional breeding projects have been significant in propelling the global aquaculture sector ahead (Gjedrem and Robinson 2014). Certain aquaculture species have shown a marked acceleration in their genetic progression as a result of the implementation of molecular technology into current breeding operations (Yue 2014). In recent years, marker-assisted selection (MAS) has been used to enhance resistance to diseases in fishes, such as Salmon resistant to IPN (Houston *et al.*, 2008), Japanese flounder resilience to lymphocystis (Fuji *et al.*, 2007), and Tilapia's potential to produce monosex population exclusively (Chen *et al.*, 2019). Additional biotechnological innovations consisting of sex control, polyploidization, gynogenesis and androgenesis have contributed a significant part in boosting aquaculture production (Zhou and Gui 2018).

Genomic selection (GS) is a revolutionary molecular breeding technique (Meuwissen *et al.*, 2001). It incorporates multiple markers as indicators for efficiency and hence, provides accurate estimates of breeding efficacy. A diverse group of aquaculture species are progressively utilizing GS (single-nucleotide polymorphism) genotyping, which uses SNPs, involving the entire genome and/or specific SNPs associated with traits, to facilitate selective breeding and stimulate genetic improvement. This is due to the ongoing advancements in sequencing and bioinformatics innovations as well as the decreasing expense of SNP genotyping (Shen and yue 2019). Previous investigations provide additional insights on GS (Houston *et al.*, 2020; Zenger *et al.*, 2019). Whenever the genes needed to be altered are acknowledged, genome editing (GE) with CRISPR/Cas can expedite the genetic modification of aquaculture species (Gratacap *et al.*, 2019). With the use of GE, beneficial alleles may be swiftly transferred to the genome, the prevalence of desirable alleles at the loci regulating significant characteristics can be increased, novel alleles can be produced, and advantageous alleles from different species can be introduced (Shen and yue 2019). Since, aquaculture animals are highly fecund and externally fertilized, enabling contemporaneously genome editing of several individuals, and hence, they are particularly well suited for genetic engineering. Reviews and books by Luo, 2019 that have been published can provide readers with comprehensive methods to GE and possible obstacles (Gratacap *et al.*, 2019; Luo 2019). A key problem in genetic engineering of aquaculture species is identifying the appropriate genes that can be modified efficiently in order to alter their traits. The appropriate genes for altering can potentially be selected with the use of beneficial details on gene functions in humans, animals, model organisms, and well-known aquaculture species. The aquaculture industry is soon going to undergo substantial changes owing to advancements in GS and GE that will enhance several aquaculture species's commercially significant features. Aquaculture's genetic advancement will be further accelerated in the future by integrating GS and GE with proficient conventional breeding techniques and advanced biotechnology related techniques. It is hoped that customers will comprehend the potential and drawbacks of genetic engineering and embrace these cutting-edge innovations for genetic advancement in the aquaculture sector. .

5. Surrogate Broodstock Technology

Selective breeding programs that are managed efectively have been proven effective in improving productive traits over time for several high-value fish species. These programs are usually enhanced by the use of genomic technologies. Gjedrem and Rye (2018) observe that these outcomes have been accomplished by progressive improvements in production traits (Gjedrem and Rye 2018). Nevertheless, the heritability of the desired traits, the species' generational gap, and the necessity to concurrently target several traits in the breeding goal restrict the benefits that may be achieved through selective breeding. Furthermore, modern breeding techniques are usually closed networks that merely encompass new variation resulting from de novo mutations and existing genetic variation in the broodstock that is usually derived from a restricted sample of wild populations. By implanting donor animal germ cells into sterile beneficiaries (the surrogates), who might belong to members of the same species or an analogous one, produces gametes derived from donors in surrogate parents (Yoshizaki and Yazawa 2019). The effective inter-species germ cell transfer, fertilization, and development of viable offspring have all been demonstrated by the significant applications of this technique in a numner of high-value aquaculture species parents (Yoshizaki and Yazawa 2019). The use of surrogate broodstock technology in aquaculture has not yet reached its full potential, but it has immense possibilities as a research tool and as an approach of advancing selective breeding.

The most astounding feature of this approach is that females being receivers, develop productive eggs that are produced from the donor's cells following the infusion of spermatogonial stem cells lined up from the donor's testis (Okutsu *et al.*, 2006) alongside male recipients develop effective sperm acquired from the donor subsequent to the transplantation of oogonial stem cells lined up from the donor ovaries (Yoshizaki *et al.*, 2010). Apparently, spermatogonial stem cells are likely to produce sperm and oogonial stem cells would produce eggs if the sexes of both the donor and the recipient coincide. Therefore, donor germline stem cells differentiate their gametes in accordance to the recipient's gender instead of depending on their own sexual orientation (Yoshizaki *et al.*, 2010b). In cases where rainbow trout cells had been transplanted into a strain of a wild-type masu salmon, the masu salmon recipients concurrently formed gametes representing both rainbow trout and masu salmon (Okutsu *et al.*, 2008b). Furthermore, by transferring diploid rainbow trout germline stem cells into triploid sterile masu salmon, it appeared feasible to develop a masu salmon that solely developed rainbow trout gametes (Okutsu *et al.*, 2007). Stem cell suppression or knockout of the dead end (dnd) gene, that's essential for the continual proliferation of primordial germ cells, is potentially utilized to develop sterile recipients devoid of endogenous germ cells as a possible substitute to triploidization (Li *et al.*, 2017; Octavera and Yoshizaki 2019). Several species of fish capable of developing smaller eggs, including the yellowtail Seriola quinqueradiata (Morita *et al.*, 2015), the tiger puffer Takifugu rubripes (Hamasaki *et al.*, 2017), the nibe croaker Nibea mitsukurii (Yoshikawa *et al.*, 2017), and the chub mackerel Scomber japonicus (Tani *et al.*, unpublished data) have been produced effectively employing the aforementioned intra-peritoneal transplantation of germline stem cells. Therefore, as long as the fish are oviparous, this technique could potentially be used to a variety of economically viable fish species. However, the issue regarding the extent to which genetic differences among donor and recipient are permissible must be taken into account when evaluating the implementation of this technique. Though additional studies needs to be conducted to better understand the limitations of genetic distance, based on the earlier studies recipients are capable to synthesize donor-derived eggs and sperm as long as they are in a similar genera irrespective of whether they belong to distinct species (Bar *et al.*, 2016). Most recently, researchers additionally determined that species belonging to the Acheilognathinae (Octavera *et al.*, unpublished data) and Salmonidae (Fujihara *et al.*, unpublished data) had achieved successful production of donor-derived eggs employing inter-generic transplantation.

6. Chromosome Engineering

Technologies for chromosomal sex modification have been widely used in farmed fish species to produce polyploidy i.e., triploidy and tetraploidy and uniparental chromosome inheritance i.e., primarily gynogenesis and androgenesis (Pandian and Sheela 1995; Lakra and Das 1998). In accordance with the potential to expedite gonadal sterilization, sex control hybrid viability development and cloning procedures have become fundamental in the evolution of fish breeding. With regard to aquaculture and fisheries management, induced triploidy is generally acknowledged as the most successful technique to develop sterile fish (Bonger *et al.*, 1994). Individuals with polyploidy possess more than one extra sets of chromosomes; triploids contain three sets, tetraploids include four, and further on. The utilization of tetraploid breeding lines in aquaculture appears to be feasible as they provide an efficient method of producing significant quantities of sterile triploid fish via relatively simple interploidy crossings among tetraploids and diploids (Chourrout *et al.*, 1986). Despite the fact that tetraploidy has been successfully induced in numerous finfish species, the survival of tetraploids was restricted in the majority of cases. The techniques of triploidy induction include subjecting fertilized eggs to shocking temperatures (hot or cold), hydrostatic pressure shock or drugs such a colchicines, cytochalasin-B or nitrous oxide in order to inhibit the initial mitotic division (Gomelsky *et al.*, 2000). Tetraploid induction includes fertilizing eggs together with normal sperm and subjecting the diploid zygote to physical or chemical-based treatment. The technique of animal development that is exclusively inherited from mother is called gynogenesis. Fish breeders are particularly interested in producing gynogenetic individuals since they may produce a substantial level of inbreeding in a single generation. In species where female homogamety exists, gynogenesis can additionally be utilized to generate exclusively female populations to uncover the processes behind fish sex determination. With regard to sex inversion investigations, it is more practical to employ entirely female gynogenetic progenies rather than typical bisexual progenies. In the majority of aquaculture species the use of hormonal sex inversion and enforced gynogenesis have been devised (Gomelsky *et al.*, 2000). Androgenesis is a technique that has economic applications in aquaculture. Moreover, it could potentially utilize to develop homozygous fish lines and retrieve repressed genotypes using cryopreserved sperm.

7. Aquatic Disease Management

Numerous vaccines towards bacteria and viruses have been produced specifically to assist with aquaculture of finfish (FAO 1996). Certain vaccinations were formerly traditional ones composed of dead microorganisms, however, innovative vaccines comprised of genetically modified organisms, DNA, and protein subunits have recently been in progression. One of the main obstacles to aquaculture's advancement is the prevalence of disease. Exclusion of viruses is particularly significant in developing resistance to disease in fish and shellfish species. Biotechnological approaches like molecular diagnostic methods, vaccinations, and immunostimulants are becoming ever more prominent in this regard (Bonger *et al.*, 1994; Carolfeld *et al.*, 2000). In these circumstances, a quick approach for pathogen identification is required. This emerging discipline is currently experiencing significant advancements in the use of biotechnological tools including polymerase chain reaction (PCR) and gene probes. Gene probes and PCR assisted diagnostics approaches have been explored for a variety of infections impacting fish and shrimp (Karunasagar and Karunasagar 1999). A prevalent tactic in the vertebrate systems is vaccination towards an illness. Despite shrimps' relatively underdeveloped immune systems, biotechnology technologies have proven useful in the production of molecules that can strengthen their defenses. Researches conducted recently have demonstrated that microbe-derived substances including lipopolysaccharides, peptidoglycans, or glucans have the potential to enhance the nonspecific defense system (Sakai 1999). Levomisole and glucan are two immunostimulants that have been found to be beneficial for fish, since they both stimulate phagocytic activity along with specific antibody responses (Sakai 1999). Current methods for characterizing various pathogen species and strains include DNA-based technology. Genetic characterization of the disease may potentially offer insight into its origin e.g. two distinct strains of the crayfish plague fungus, one from the native species along with another from Turkey, were found in Sweden based on DNA sequencing (FAO/NACA/CSIRO/ACIAR/DFID 1999). Afterwards the pathogen has been determined, DNA probes can potentially be designed to validate against specific pathogens in tissue, in water and soil specimens. According to Subasinghe (2009) and Subasinghe & Bondad-Reantaso (2006), these methods have been implemented to identify bacterial and fungal infections in fish as well as viral infections affecting marine shrimp worldwide (Subasinghe 2009; Subasinghe and Bondad-Reantaso 2006). Acquiring fast, dependable, and highly precise screening assays is necessary for the efficient control and management of aquatic organism diseases. Initiatives to address these issues might be successful with the introduction of immunoassays, DNA-based diagnostic, and polymerase chain reaction (PCR) amplification technologies.

The process of vaccination involves exposing the host's cells to biological compounds that enable the host to develop a specific immune response. This response provides the host an advantage over genetically comparable non-vaccinated hosts in terms of combating off subsequent infections by targeting a specific pathogen. Additionally, it has been demonstrated to be economical and contributed to reduction in the consumption of antibiotics. Immunostimulants and vaccinations can be introduced as immersion, or may be directly added into meals and for larger aquatic animals may be delivered via injections. According to Subasinghe (2009), fish vaccinations, which have been produced in the past couple of decades, are acknowledged, tested, and affordable approach to prevent several infectious diseases in cultured animals across the globe (Subasinghe 2009). Other commercialized vaccines have long been accessible for finfish infections, such as furunculosis (Aeromonas salmonicida), and even more are currently under consideration, such as viral hemorrhagic septicemia (VHS). Vaccines not only minimize the prevalence of disease-related deaths but also forestall the demand for antibiotics, do not leave traces in the natural environment or on products, and do not induce resistance to pathogenic organisms (Subasinghe and Bondad-Reantaso 2006).

The vaccination approach for fish involves subjecting the fish's immune system to either the whole or a portion of the pathogen. After a specific amount of time, immunity builds. Fish vaccines are divided into two categories: killed fish vaccines and modified live vaccinations. Vaccinations that have been formalin-or heat-killed are known as killed fish vaccines. According to Duff's 1942 study, the first documented application of a vaccination in fish was a killed vaccine for pathogenic strain of Aeromonas salmonicida (Duff 1942). Ma *et al.* (2019) worked on oral vaccination in Oncorhynchus clarkia (Ma *et al.*, 2019). In contrast to killed preparations, modified live vaccines are typically made up of microorganisms those are alive and therefore more likely to produce an immune response because of their tendency for proliferation, effortless penetration into the host, and a higher degree of cellular response stimulation with respect to both intrinsic and adaptive immunities (Levine and Sztein 2004). Approximately three modified live vaccines are reportedly approved in the United States at present. These vaccines are, Arthrobacter vaccine for salmonids against bacterial kidney disease (BKD), Flavobacterium columnare vaccine for catfish against columnaris infection, and Edwardsiella ictaluri vaccine against enteric septicemia infection in catfishes (Klesius and Pridgeon 2014).

Vaccines that have been live-attenuated operate by inducing humoral and cell-mediated immunity. Live vaccinations have drawn criticism, meanwhile, because of concerns about environmental sustainability. In fish such as grass carp, immunoprophylactic methods have been identified to activate indicators for virus nucleic acids, toll-like receptors (TLRs), high mobility group box proteins (HMGBs), retinoic acid inducible gene-I (RIG-I)–like receptors (RLRs), and pattern recognition receptors (PRRs). According to modern molecular research, they can be regulated to produce the appropriate ranges of immunity and are important in stimulating the fish body's defenses in combating viral infections (Rao and Su 2015). Their significance in adaptive immunity has also been reported. According to Rauta *et al.* (2014), a successful vaccination for fish and other aquatic species may be developed via the incorporation of TLRs as an adjuvant and TLR activators in the vaccine formulations (Rauta *et al.* 2014). The innovations in the designing and development of vaccines, such as immunomics-based vaccines, designer cell lines, dendritic cells, marker vaccines, and structural vaccinology (SV), is an additional current breakthrough in the field of molecular biology (Delany *et al.* 2014; Singh *et al.* 2015). According to Perez *et al.* (2013), adjuvants that have been improved may boost the degree of defense in the required level, which can be beneficial in facilitating the development of vaccines for circumstances when vaccines do not elicit significant immune responses (Malik *et al.* 2015).

7.1 Reverse Vaccinology

Reverse vaccinology is a modern technique that has gained attention owing to an upsurge in biotechnology advancements. A reliable and effective vaccination takes generations to formulate, according to Rappuoli (2000). The development of vaccines has been scaled up from five to ten years to 1-2 years with the use of this present vaccinology paradigm. Dadar *et al.* (2016) stated that the application of bioinformatics may reduce the time drastically to generate new vaccines rather than generating new vaccines via traditional methods. Photobacterium damselae subsp. piscicida is one of the marine species to which this technique has employedrecently. Mahendran *et al.*, (2016) reported that software to develop vaccines for Flavobacterium columnare and Edwardsiella tarda, the two significant intracellular fish infections that cause columnaris and edwardsiellosis, respectively has been studied. There are also severe drawbacks associated with reverse vaccination, as demonstrated by several researches (Rappuoli 2000; Dadar *et al.*, 2016). (Wright some of the drawbacks if not able to find then delete the statement otherwise the paragraph is not continuous). Ellul *et al.*, research report from 2021 emphasized the possibilities of using Cyclopterus lumpus reverse vaccinology against Pasteurella atlantica to avoid pasteurellosis in aquaculture. In order to stop disease outbreaks, the most robust gene target candidates should be provided preference in the production of vaccines as reported by their insilco functional analysis study.

8. Applications of Nanotechnology in Aquaculture

Numerous studies have been conducted on the application of nanoparticles in the processing of aquaculture effluents; nevertheless, there remain limitations concerning issues regarding expenses, energy consumption for synthesizing, toxicities, and the handling of waste from the technology in this context. Future research demand to focus on developing improved techniques to enhance the characteristics of nanomaterials utilized for such applications (Ellul *et al.*, 2021). Aquaculture farmers find it unappealing since nanotechnology procedures are more expensive as opposed to conventional waste treatments. Hence, study should be undertaken regarding the application of low-cost, natural techniques and materials that might reduce the expense of nanotechnology for the treatment of waste water. In this context, cost effective plant-mediated nanoparticle production may be employed to remediate these effluents (Sahu *et al.*, 2021). Further studies should be conducted to improve the effectiveness of wastewater treatment & lessen the toxicity of nanoparticles and nanoclays on fish and the aquatic environment. These two methods of treating wastewater include adsorption, but they also have the potential to remove organic contaminants, anions, and heavy metals commonly found in aquaculture effluents. Consideration needs to be focused on treating aquaculture with less hazardous nanomaterials that have beneficial qualities, such as gold nanorods doped poly(styrenesulfonate) composites. Furthermore, nanomaterials whose toxicity are size dependant such as titanium oxide nanoparticles, and aluminum oxide with carbon ABB black composite should be preferred. Yaqoob *et al.* (2020) suggest the compounds those toxicity is concentration-dependent, like Ag/Carbon composite, single- and multi-walled carbon nanotubes, and Hexadecylcetyl trimethyl ammonium bromide (CTAB) doped Gold nanorod composites, ought to be further studied for their potential use in treating aquaculture effluents and potential toxicity towards fish and other aquatic life (Sahu *et al.*, 2021). There is a lot of possibilities in the field of photocatalysis for treating such effluent. According to Cai *et al.*, (2018), nano-sized photocatalysts, such as nanotitanium oxide, have been recognized for their superior efficiency and reduced cost when compared to certain other nano-materials (Yaqoob *et al.*, 2020; Cai *et al.*, 2018). Their potential toxicity to fish and other aquatic species should be further investigated. Utilizing plants as bio-filters to remediate wastewater is a rapidly developing area. According to Turcios and Papenbrock (2014), plants including water lettuce, water hyacinth, and Salicornia dolychosta, Moss, a member of the Salicornia genus, have become more prominently recognized for their usage as biofilters; (Cai *et al.*, 2018). These plants having the capability to function as biofilters can be enhanced and their growth and biomass may be amplified with the use of nanotechnology making them more fish-friendly. The reason for using nanotechnology is establishing an environmentally friendly aquaculture environment and goes beyond the simple desire to produce larger quantities of food and proteins; which will provide a sustainable environment free from industrial operations.

9. Biosecurity and Disease Control

Like other agricultural practices, the aquaculture sector has experienced a significant number of transboundary aquatic animal diseases brought about by bacteria, fungi, viruses, parasites, and other pathogens that are however, unknown and increasing in abundance. Rapid and precise diagnostic technologies are important for controlling pandemics since they facilitate the identification and detection of the pathogen responsible for mortality. Aquaculture field has made considerable use of DNA and RNA techniques for the detection of various bacterial and viral diseases globally. The methods are based on the unique DNA or RNA sequences that are carried by every pathogenic species and may be utilized in their recognition. With direct applicability and fast screening for several hazardous bacterial and viral disorders made possible by the commercial accessible PCR primers and diagnostic kits, the approaches provide great accuracy and precision. When an animal demonstrates no antibody reaction following infection, molecular methods like polymerase chain reaction (PCR) could potentially be effective. As an illustration, pathogen identification in molluscs is restricted by the lack of antibody production in these organisms, making antibody-based diagnostic techniques ineffective.

Many molecular and antibody-based tests are being proposed to identify fish infections mostly caused by viruses and bacteria, whereas screening for parasites and fungi have also been described recently. A few of the tests that use antibodies involve slide agglutination, co-agglutination/latex agglutination, immunodiffusion, fluorescent antibody assays (direct and indirect), immunohistochemistry, ELISA, dot blot/dip-stick, and Western blot.

There are several parameters that determine whether antibody-based test is the most effective in identifying infections, as each approach has positive impacts and drawbacks. These approaches have limited applicability in environmental specimens, particularly in cases when pathogen concentrations are very low, despite being helpful for detecting infections in pure culture or/and infectious fish tissue. ISH and PCR, on the other hand, are excellent techniques for detecting DNA at very low concentration.

In epidemiological studies, molecular techniques may be applied to identify individual strains and distinguish closely associated strains, as well as to identify diseases down to the level of the species (Turcios and Papenbrock 2014; Puttinaowarat *et al.*, 2000). As a result of the general inadequacy of the traditional pathogen isolating procedures and immunodiagnostics for molluscs and crustaceans, molecular approaches are progressively being employed (Cawley *et al.*, 1999; Lightner 1998). PCR and other DNA-based techniques are incredibly precise. Although because of inhibition or contamination, results that are falsely positive or negative might contribute complications during diagnosis (Morris *et al.*, 2002).

Alternatively, nucleic acid sequence-based amplification and real-time PCR (closed tube to minimize contamination) provide significant sample throughput while lowering these uncertainties (Overturf *et al.*, 2001; Starkey *et al.*, 2004). Nested PCR, RAPDs, reverse transcriptase PCR (RT-PCR), reverse cross blot PCR, and RT-PCR enzyme hybridization test constitute several of the most commonly employed PCR-based pathogen identification approaches (Puttinaowarat *et al.*, 2000; Cunningham, 2004). Additionally, In situ hybridization (ISH) has also been utilized to validate the presence of mollusk parasites (Cochennec *et al.*, 2000; Carnegie *et al.*, 2003) as well as to identify shrimp viruses (Lightner 1998; Tang and Lightner 1999). According to Powell and Loutit (2004), colony hybridization possesses the benefit of being able to recognize both pathogenic and environmental strains of Vibrio anguillarum, and it has also been utilized effectively for the quick identification of this organism in fish (Aoki *et al.*, 1989). Techniques centered around RNA and DNA have been developed to identify genetic material from pathogens. The nested PCR method, which amplifies DNA further using primers from the first step amplified in the process, is a more sensitive alternative to traditional PCR with respect to the pathogen. RT-PCR may be used to quantify the amount of viral nucleic acids in sample tissues in terms of RNA. It must also be emphasized that, like the previously discussed immunological techniques, PCR only identifies the genetic material of pathogens in the sample under investigation—it does not reveal the existence of disease or a living/dead pathogen. Meanwhile, PCR provides a reasonably quick and affordable method for routinely screening a large number of aquatic animals for the purpose of commercial aquaculture and for evaluating imported stocks while they are under isolation. Presently, PCR is crucial for the periodic examination of large populations of penaeid shrimp larvae in Asian and Latin American nations for major viral diseases like WSSV, TSV, and others.

10. Genetically Modified Organisms (Transgenics)

Selective gene transfer for desirable qualities from a particular breed to another is known as transgenics. These characteristics might involve enhancement of growth rates, greater size, increased effectiveness in transformation of feed into muscle and management of sexual maturity (Hew *et al.*, 1992). Zhu *et al.*, (1985) developed the first transgenic fish in China although the molecular proof for the transgene's incorporation was lacking & the expression of putative transgenics was transitory. Numerous fish species, including salmon, trout, and tilapia, possess growth hormone genes coming from animals effectively inserted into them, leading to growth that is multiple times quicker than that of the native species. This process of creating novel varieties is quicker and more precise than other conventional method e.g., selection, hybridization, chromosome manipulation etc. The exchange of genes amongst species is another option if conventional breeding is unable to provide the necessary features. To acquire specific features and generate an effective biological product, such as the introduction of the antifreeze protein gene (AFP) in fish to acclimatized into to a freezing environment (Hew *et al.*, 1992). Since disease management is the key to aquaculture's intensification and sustainable growth, biotechnology is crucial for enhancing pathogen resistance and improving the well-being of farmed species via disease resistance selection (El-Zaeem and Aseem 2004). Additionally, disease resistance strain can be developed by using this technology. Developments in embryonic stem cell (ESC) technology are unquestionably the most potential tool for the advancement of transgenic fish breeding. Due to their totipotent status and lack of differentiation, these cells can be altered in vitro and then reintroduced into early embryos to contribute to the host's germ line. This would enable the genes to be effectively inserted or removed (Melamed *et al.*, 2002).

11. Metabolomics in Aquaculture

The successful growth of the fish industry has been largely attributed to the discovery of vaccinations against various diseases (Ringø *et al.*, 2014), but viable therapies against infections for other organisms, such as viral infections in shrimp (Seibert and Pinto 2012), are still lacking. Additional research is needed in the domain of metabolomics-based techniques to improve immunodiagnostics and host resistance, as well as to comprehend the causes of disease susceptibility. Moreover, there are significant differences in immunological response to pathogen exposure both within and between species. As demonstrated by recent studies (Nato *et al.*, 2014 & Gray *et al.*, 2015), metabolomics is a highly valuable application for studying immune mechanisms and the physiological effects of pathogen exposure in mammals. It is possible to employ this technique more broadly to study cultured aquatic species. Future applications of metabolomics may be found in the fields of selective breeding, cryopreservation, and climate change research. The hazards of global warming and ocean acidification are gradually becoming an important issue for aquaculture producers and stakeholders (Reid and Jackson 2014). Variations in the pH and sea surface temperature of the water can modify primary productivity, increase the frequency of noxious algal blooms, and increase the incidence of marine infectious diseases (Gilbert *et al.*, 2014; Richards *et al.*, 2015 & Richards *et al.*, 2015). In recent years, metabolomics has been effectively utilized in investigating the consequences of heat stress on mussels (Ellis *et al.*, 2014), sea cucumbers (Shao *et al.*, 2015), and oysters (Wei *et al.*, 2015). It has also been used to study the connections among red tide-forming dinoflagellates alongside various phytoplankton species (Poulson-Ellestad *et al.*, 2014). Genetic material can be preserved for subsequent use by cryopreservation of gametes and embryos but cryopreservation of oocytes continues to present significant challenges (Hassan *et al.*, 2015). It is possible to cryopreserve fertilized eggs in molluscs, however, the causes for the typically exceedingly poor probability of survival are unclear. Therefore, metabolomics may be utilized to improve our comprehension of the physiological mechanisms causing cellular damage during the freezing and warming processes (Abuja *et al.*, 2015; Koštál *et al.*, 2015). It may also aid in the development of repair techniques for cryoinjury. Metabolomics can also be applied in the field of selective breeding,recently,. Pushpa *et al.* (2014), Kushalappa & Gunnaiah (2013), and Buitenhuis *et al.*, (2013) have reported the discovery of novel phenotypes by applying metabolomics in agriculture. Therfore, metabolomics has tremendous potential to be utilized in selective breeding of aquatic animals to improve their productivity, resistance to disease, and overall quality. This offers a fascinating new study direction that may be used to enhance and expand the aquaculture industry.

Table 1. Summary of novel emerging technologies in the aquaculture sector

Program	**Technique**	**Fish Species**	**Application**	**References**
Fish breeding	Induced breeding	Perch, Catfish, Scallop	To achieve maturation & breeding in captivity	(Halder, *et al.*, 1991; Tang, *et al.*, 2023)
Aquaculture nutrition	Fish feed development	Salmon	To enhance protein concentration	(Hodar, *et al.*, 2020; Han, *et al.*, 2019; Jones, *et al.*, 2020 & Sharif, *et al.*, 2021).
Selective breeding	Genomic selection	Salmon, Tilapia	To improve the breeding efficacy & control diseases	(Houston, *et al.*, 2008; Meuwissen, *et al.*, 2008 & Shen and yue 2019)
Fish Species Selection	Surrogate broodstock technology	Masu salmon, Rainbow trout, Yellowtail, Tiger puffer, Nibe croaker, Chub mackerel	To improve productive traits	(Yoshizaki, and Yazawa 2019; Okutsu, *et al.*, 2008; Morita, *et al.*, 2015 Hamasaki, *et al.*, 2015; Yoshikawa, *et al.*, 2017
Fish Sex modification	Chromosome Engineering	Salmon, Tilapia	To produce polyploidy	(Pandian and Sheela 1995; Lakra and Das 1998)
Fish Health Management	Molecular diagnostic methods, Vaccinations, Reverse vaccinology and Immunostimulants	Shrimp, Prawn, Crayfish	To develop resistance against diseases	(Karunasagar and Karunasagar, 1999; Carolsfeld, *et al.*, 2000; Baonger, *et al.*, 1994; Subasinghe 2009 & Subasinghe and Bonad-Reamtaso 2006)
Aquaculture waste Management	Nanotechnology	Salicornia spp.	To treat aquaculture effluent pollution	(Sahu, *et al.*, 2021; Yaqoob, *et al.*, 2020 & Cai, *et al.*, 2018)
Bio-security and disease Management	PCR, RAPDs & Diagnostic kit	Shrimp	To control pandemics & identification of pathogen	(Turcios and Papenbrock 2014; Lightner 1998; Morris, *et al.*, 2002; Tang and Lightner 1999)
Fish Gene Modification	Transgenic & Embryo stem cell technology	Salmon (Aquadvantage), Tilapia, Zebrafish	To improve fish health, breeding and sexual maturity	(Zhu, *et al.*, 1985; Hew, *et al.*,1992 & Melamed, *et al.*, 2002)
Metabolomics	(Selective breeding, Cryopreservation)	Shrimp, Mussels, & Oysters	For effective management of diseases, Heat stress	(Ringø, *et al.*, 2014; Seibert and Pinto 2012; Reid and Jackson 2014; Ellis, *et al.*, 2014; Shao, *et al.*, 2015; Wei, *et al.*, 2015 & Poulson-Ellestad, *et al.*, 2014)

Conclusion

In the last few decades, biotechnology has become increasingly significant in the advancement of the well-being of people, the agricultural sector, and aquaculture. We are now equipped with novel techniques and have incredible capacity to develop unique genotypes and genes in fish, plants, and animals. Utilizing biotechnology in aquaculture industry is a relatively new endeavor, yet there is a huge potential to increase the fish production to meet the ever - increasing demand and completely transform aquaculture. The study provides a brief overview of recent developments and focus areas in the fields of transgenesis, aquaculture nutrition, chromosomal engineering, artificial hormone usage in fish breeding, biotechnology for health care, and gene banking. Governmental assistance is crucial for boosting aquaculture growth without having an adverse effect on the environment since, it removes ambiguities and susceptibility in the risks. The bottom lines of sustainable production, economic viability, prosperity, environmental protection, social security and acceptability will surely be further satisfied by improved information sharing, cooperation, collaboration, and synergy among researchers and producers on their issues and prospects.

References

Abuja, P. M., Ehrhart, F., Schoen, U., Schmidt, T., Stracke, F., Dallmann, G., Friedrich, T., Zimmermann, H., & Zatloukal, K. (2015). Alterations in Human Liver Metabolome during Prolonged Cryostorage. Journal of Proteome Research, 14(7), 2758–2768. https://doi.org/10.1021/acs.jproteome.5b00025

Adams, A., & Thompson, K. D. (2008). Recent applications of biotechnology to novel diagnostics for aquatic animals. Revue Scientifique Et Technique (International Office of Epizootics), 27(1), 197–209. https://pubmed.ncbi.nlm.nih.gov/18666488/

Adelizi, P. (1998). Agbiotech Information Bulletin for Schools. Issue 33. Published by AG-WEST BIOTECH INC. Sask., S7N 3R2 Canada.

Andreoni, F., Amagliani, G., & Magnani, M. (2016). Selection of Vaccine Candidates for Fish Pasteurellosis Using Reverse Vaccinology and an In Vitro Screening Approach. In S. Thomas (Ed.), Vaccine Design (Vol. 1404, pp. 181–192). Springer New York. https://doi.org/10.1007/978-1-4939-3389-1_12

Aoki, T., Hirono, I., Castro, T., & Kitao, T. (1989). Rapid identification of Vibrio anguillarum by Colony hybridization. Journal of Applied Ichthyology, 5(2), 67–73. https://doi.org/10.1111/j.1439-0426.1989.tb00475.x

Bar, I., Smith, A., Bubner, E., Yoshizaki, G., Takeuchi, Y., Yazawa, R., Chen, B. N., Cummins, S., & Elizur, A. (2016). Assessment of yellowtail kingfish (Seriola lalandi) as a surrogate host for the production of southern bluefin tuna (Thunnus maccoyii) seed via spermatogonial germ cell transplantation. Reproduction, Fertility and Development, 28(12), 2051. https://doi.org/10.1071/RD15136

Beck, B. R., Kim, D., Jeon, J., Lee, S.-M., Kim, H. K., Kim, O.-J., Lee, J. I., Suh, B. S., Do, H. K., Lee, K. H., Holzapfel, W. H., Hwang, J. Y., Kwon, M. G., & Song, S. K. (2015). The effects of combined dietary probiotics Lactococcus lactis BFE920 and Lactobacillus plantarum FGL0001 on innate immunity and disease resistance in olive flounder

(Paralichthys olivaceus). Fish & Shellfish Immunology, 42(1), 177–183. https://doi.org/10.1016/j.fsi.2014.10.035

Bhattacharya, S., Dasgupta, S., Datta, M. and Basu, D. 2002. Biotechnology Input in Fish Breeding. Indian J. Biotechnol., 1: 29-38. https://nopr.niscair.res.in/bitstream/123456789/19848/1/IJBT%201(1)%2029-38.pdf

Bonger, A. B. J., Abo-Hashema, K., Bremmer, I. M., Eding, E. H., Komen, J. & Richter, C. J. J. (1994). Androgenesis in common carp (Cyprinus carpio) using UV Irradiation in a synthetic ovarian fluid and heat shock, aquaculture, 122, 199-132. https://research.wur.nl/en/publications/androgenesis-in-common-carp-cyprinus-carpio-l-using-uv-irradiatio

Buitenhuis, A. J., Sundekilde, U. K., Poulsen, N. A., Bertram, H. C., Larsen, L. B., & Sørensen, P. (2013). Estimation of genetic parameters and detection of quantitative trait loci for metabolites in Danish Holstein milk. Journal of Dairy Science, 96(5), 3285–3295. https://doi.org/10.3168/jds.2012-5914

Cai, Z., Dwivedi, A. D., Lee, W.-N., Zhao, X., Liu, W., Sillanpää, M., Zhao, D., Huang, C.-H., & Fu, J. (2018). Application of nanotechnologies for removing pharmaceutically active compounds from water: Development and future trends. Environmental Science: Nano, 5(1), 27–47. https://doi.org/10.1039/C7EN00644F

Cao, L., Naylor, R., Henriksson, P., Leadbitter, D., Metian, M., Troell, M., & Zhang, W. (2015). China's aquaculture and the world's wild fisheries. Science, 347(6218), 133–135. https://doi.org/10.1126/science.1260149

Carnegie, R., Meyer, G., Blackbourn, J., Cochennec-Laureau, N., Berthe, F., & Bower, S. (2003). Molecular detection of the oyster parasite Mikrocytos mackini, and a preliminary phylogenetic analysis. Diseases of Aquatic Organisms, 54, 219–227. https://doi.org/10.3354/dao054219

Carolsfeld, J., Powell, J. F. F., Park, M., Fischer, W. H., Craig, A. G., Chang, J. P., Rivier, J. E., & Sherwood, N. M. (2000). Primary Structure and Function of Three Gonadotropin-Releasing Hormones, Including a Novel Form, from an Ancient Teleost, Herring 1. Endocrinology, 141(2), 505–512. https://doi.org/10.1210/endo.141.2.7300

Chen, C.H., Li, B.J., Gu, X.H., Lin, H.R., & Xia, J. H. (2019). Marker-assisted selection of YY supermales from a genetically improved farmed tilapia-derived strain. Zool Res. 2019 Mar 18; 40(2):108-112. https://doi.org/10.24272/j.issn.2095-8137.2018.071

Chourrout, D., Chevassus, B., Krieg, F., Happe, A., Burger, G., & Renard, P. (1986). Production of second generation triploid and tetraploid rainbow trout by mating tetraploid males and diploid females ? Potential of tetraploid fish. Theoretical and Applied Genetics, 72(2), 193–206. https://doi.org/10.1007/BF00266992

Cochennec, N., Le Roux, F., Berthe, F., & Gerard, A. (2000). Detection of Bonamia ostreae Based on Small Subunit Ribosomal Probe. Journal of Invertebrate Pathology, 76(1), 26–32. https://doi.org/10.1006/jipa.2000.4939

Conceição, L. E. C., Yúfera, M., Makridis, P., Morais, S., & Dinis, M. T. (2010). Live feeds for early stages of fish rearing. Aquaculture Research, 41(5), 613–640. https://doi.org/10.1111/j.1365-2109.2009.02242.x

Cowley, J., Dimmock, C., Wongteerasupaya, C., Boonsaeng, V., Panyim, S., & Walker, P. (1999). Yellow head virus from Thailand and gill-associated virus from Australia are closely related but distinct prawn viruses. Diseases of Aquatic Organisms, 36, 153–157. https://doi.org/10.3354/dao036153

Cunningham, C. O. (2004). Use of Molecular Diagnostic Tests in Disease Control: Making the Leap from Laboratory Development to Field Application. In L. K. Yin, Molecular Aspects of Fish & Marine Biology (Vol. 3, pp. 292–312). WORLD SCIENTIFIC. https://doi.org/10.1142/9789812565709_0011

Dadar, M., Dhama, K., Vakharia, V. N., Hoseinifar, S. H., Karthik, K., Tiwari, R., Khandia, R., Munjal, A., Salgado-Miranda, C., & Joshi, S. K. (2017). Advances in Aquaculture Vaccines Against Fish Pathogens: Global Status and Current Trends. Reviews in Fisheries Science & Aquaculture, 25(3), 184–217. https://doi.org/10.1080/23308249.2016.1261277

Delany, I., Rappuoli, R., & De Gregorio, E. (2014). Vaccines for the 21st century. EMBO Molecular Medicine, 6(6), 708–720. https://doi.org/10.1002/emmm.201403876

Duff, D. C. B. (1942). The Oral Immunization of Trout Against Bacterium Salmonicida. The Journal of Immunology, 44(1), 87–94. https://doi.org/10.4049/jimmunol.44.1.87

El-Zaeem, S. Y., & Aseem, S. S. (2004). Application of Biotechnology in Fish Breeding: Production of Highly Immune Genetically Modified Nile Tilapia, Orechromis niloticus, with Accelerated Growth by Direct Injection of Shark DNA into Skeletal Muscles. Egyptian Journal of Aquatic Biology and Fisheries, 8 (3), 67-92. https://www.ijcmas.com/9-6-2020/Renu%20Singh%20and%20Babita%20Rani.pdf

Ellis, R. P., Spicer, J. I., Byrne, J. J., Sommer, U., Viant, M. R., White, D. A., & Widdicombe, S. (2014). 1 H NMR Metabolomics Reveals Contrasting Response by Male and Female Mussels Exposed to Reduced Seawater pH, Increased Temperature, and a Pathogen. Environmental Science & Technology, 48(12), 7044–7052.

Ellul, R. M., Kalatzis, P. G., Frantzen, C., Haugland, G. T., Gulla, S., Colquhoun, D. J., Middelboe, M., Wergeland, H. I., & Rønneseth, A. (2021). Genomic Analysis of Pasteurella atlantica Provides Insight on Its Virulence Factors and Phylogeny and Highlights the Potential of Reverse Vaccinology in Aquaculture. Microorganisms, 9(6), 1215. https://doi.org/10.3390/microorganisms9061215

FAO(2020). The State of World Fisheries and Aquaculture 2020. Sustainability in Action. FAOhttps://doi.org/10.4060/ca9229en

Food and Agriculture Organization of the United Nations, & World Health Organization (Eds.). (1996). Biotechnology and food safety: Report of a joint FAO/WHO consultation, Rome, Italy, 30 September-4 October 1996.c. Food and Agriculture Organization of the United Nations.

Fuji, K., Hasegawa, O., Honda, K., Kumasaka, K., Sakamoto, T., & Okamoto, N. (2007). Marker-assisted breeding of a lymphocystis disease-resistant Japanese flounder (Paralichthys olivaceus). Aquaculture, 272(1–4), 291–295. https://doi.org/10.1016/j.aquaculture.2007.07.210

Gjedrem, T., & Robinson, N. (2014). Advances by Selective Breeding for Aquatic Species: A Review. Agricultural Sciences, 05(12), 1152–1158. https://doi.org/10.4236/as.2014.512125

Gjedrem, T., & Rye, M. (2018). Selection response in fish and shellfish: A review. Reviews in Aquaculture, 10(1), 168–179. https://doi.org/10.1111/raq.12154

Glibert, P. M., Icarus Allen, J., Artioli, Y., Beusen, A., Bouwman, L., Harle, J., Holmes, R., & Holt, J. (2014). Vulnerability of coastal ecosystems to changes in harmful algal bloom distribution in response to climate change: Projections based on model analysis. Global Change Biology, 20(12), 3845–3858. https://doi.org/10.1111/gcb.12662

Gomelsky, B., Mims, S. D., Onders, R. J., Shelton, W. L., Dabrowski, K., & Garcia-Abiado, M. A. (2000). Induced Gynogenesis in Black Crappie. North American Journal of Aquaculture, 62(1), 33–41. https://doi.org/10.1577/1548-8454(2000)062<0033:IGIBC>2.0.CO;2

Gratacap, R. L., Wargelius, A., Edvardsen, R. B., & Houston, R. D. (2019). Potential of Genome Editing to Improve Aquaculture Breeding and Production. Trends in Genetics, 35(9), 672–684. https://doi.org/10.1016/j.tig.2019.06.006

Gray, D. W., Welsh, M. D., Doherty, S., Mansoor, F., Chevallier, O. P., Elliott, C. T., & Mooney, M. H. (2015). Identification of systemic immune response markers through metabolomic profiling of plasma from calves given an intra-nasally delivered respiratory vaccine. Veterinary Research, 46(1), 7. https://doi.org/10.1186/s13567-014-0138-z

Halder, S., Sen, S., Bhattacharya, S., Ray, A. K., Ghosh, A., & Jhingran, A. G. (1991). Induced spawning of Indian major carps and maturation of a perch and a catfish by murrel gonadotropin releasing hormone, pimozide and calcium. Aquaculture, 97(4), 373–382. https://doi.org/10.1016/0044-8486(91)90329-6

Hamasaki, M., Takeuchi, Y., Yazawa, R., Yoshikawa, S., Kadomura, K., Yamada, T., Miyaki, K., Kikuchi, K., & Yoshizaki, G. (2017). Production of Tiger Puffer Takifugu rubripes Offspring from Triploid Grass Puffer Takifugu niphobles Parents. Marine Biotechnology, 19(6), 579–591. https://doi.org/10.1007/s10126-017-9777-1

Han, D., Shan, X., Zhang, W., Chen, Y., Wang, Q., Li, Z., Zhang, G., Xu, P., Li, J., Xie, S., Mai, K., Tang, Q., & De Silva, S. S. (2018). A revisit to fishmeal usage and associated consequences in Chinese aquaculture. Reviews in Aquaculture, 10(2), 493–507. https://doi.org/10.1111/raq.12183

Han, P., Lu, Q., Fan, L., & Zhou, W. (2019). A Review on the Use of Microalgae for Sustainable Aquaculture. Applied Sciences, 9(11), 2377. https://doi.org/10.3390/app9112377

Hassan, M. M., Qin, J. G., & Li, X. (2015). Sperm cryopreservation in oysters: A review of its current status and potentials for future application in aquaculture. Aquaculture, 438, 24–32. https://doi.org/10.1016/j.aquaculture.2014.12.037

Hew, C. L., Davies, P. L., & Fletcher, G. (1992). Antifreeze protein gene transfer in Atlantic salmon. Molecular Marine Biology and Biotechnology, 1(4–5), 309–317. https://pubmed.ncbi.nlm.nih.gov/1308821/

Hill, B. J. (2005). The need for effective disease control in international aquaculture. Developments in Biologicals, 121, 3–12. https://pubmed.ncbi.nlm.nih.gov/15962465/

Hodar, A., Vasava, R., Mahavadiya, D., & Joshi, N. (2020). Fish meal and fish oil replacement for aqua feed formulation by using alternative sources: A review. Journal of Experimental Zoology India, 23, 13–21. https://www.researchgate.net/publication/338392541_fish_meal_and_fish_oil_replacement_for_aqua_feed_formulation_by_using_alternative_sources_a_review

Houston, R. D., Bean, T. P., Macqueen, D. J., Gundappa, M. K., Jin, Y. H., Jenkins, T. L., Selly, S. L. C., Martin, S. A. M., Stevens, J. R., Santos, E. M., Davie, A., & Robledo, D. (2020). Harnessing genomics to fast-track genetic improvement in aquaculture. Nature Reviews Genetics, 21(7), 389–409. https://doi.org/10.1038/s41576-020-0227-y

Houston, R. D., Haley, C. S., Hamilton, A., Guy, D. R., Tinch, A. E., Taggart, J. B., McAndrew, B. J., & Bishop, S. C. (2008). Major Quantitative Trait Loci Affect Resistance to Infectious Pancreatic Necrosis in Atlantic Salmon (Salmo salar). Genetics, 178(2), 1109–1115. https://doi.org/10.1534/genetics.107.082974

https://books.google.co.in/ books?hl=en&lr=& id=0Qk8DQAAQBAJ & oi=fnd&

https://doi.org/10.1021/es501601w

Igwegbe, C. A., Onukwuli, O. D., Ighalo, J. O., & Umembamalu, C. J. (2021). Electrocoagulation-flocculation of aquaculture effluent using hybrid iron and aluminium electrodes: A comparative study. Chemical Engineering Journal Advances, 6, 100107. https://doi.org/10.1016/j.ceja.2021.100107

Jones, S. W., Karpol, A., Friedman, S., Maru, B. T., & Tracy, B. P. (2020). Recent advances in single cell protein use as a feed ingredient in aquaculture. Current Opinion in Biotechnology, 61, 189–197. https://doi.org/10.1016/j.copbio.2019.12.026

Karunasagar, I., & Karunasagar, I. (1999). Diagnosis, treatment and prevention of microbial diseases of fish and shellfish. Current Science, 76(3), 387–399. http://www.jstor.org/stable/24101134

Kelly, A. M., & Renukdas, N. N. (2020). Disease management of aquatic animals. In Aquaculture Health Management (pp. 137–161). Elsevier. https://doi.org/10.1016/B978-0-12-813359-0.00005-1

Klesius, P. H., & Pridgeon, J. W. (2014) Vaccination against enteric septicemia of catfish. In: Fish vaccination. John Wiley & Sons, Chichester, pp 211–225. https://www.ncbi.nlm.nih.gov/pmc/articles/PMC6920890/

Koštál, V., Zahradníčková, H., & Šimek, P. (2011). Hyperprolinemic larvae of the drosophilid fly, Chymomyza costata , survive cryopreservation in liquid nitrogen. Proceedings of the National Academy of Sciences, 108(32), 13041–13046. https://doi.org/10.1073/pnas.1107060108

Kushalappa, A. C., & Gunnaiah, R. (2013). Metabolo-proteomics to discover plant biotic stress resistance genes. Trends in Plant Science, 18(9), 522–531. https://doi.org/10.1016/j.tplants.2013.05.002

Lakra, W. S. & Das, P. (1998). Genetic engineering in aquaculture. Indian. J. Anim. Sci., 68(8), 873-879. https://epubs.icar.org.in/index.php/IJAnS/article/view/21193

Lauzon, H. L., Pérez-Sánchez, T., Merrifield, D. L., Ringø, E., & Balcázar, J. L. (2014). Probiotic Applications in Cold Water Fish Species. In D. Merrifield & E. Ringø (Eds.), Aquaculture Nutrition (1st ed., pp. 223–252). Wiley. https://doi.org/10.1002/9781118897263.ch9

Lazado, C. C., & Caipang, C. M. A. (2014). Mucosal immunity and probiotics in fish. Fish & Shellfish Immunology, 39(1), 78–89. https://doi.org/10.1016/j.fsi.2014.04.015

Levine, M. M., & Sztein, M. B. (2004). Vaccine development strategies for improving immunization: The role of modern immunology. Nature Immunology, 5(5), 460–464. https://doi.org/10.1038/ni0504-460

Levy, T., Rosen, O., Eilam, B., Azulay, D., Zohar, I., Aflalo, E. D., Benet, A., Naor, A., Shechter, A., & Sagi, A. (2017). All-female monosex culture in the freshwater prawn Macrobrachium rosenbergii – A comparative large-scale field study. Aquaculture, 479, 857–862. https://doi.org/10.1016/j.aquaculture.2017.07.039

Li, Q., Fujii, W., Naito, K., & Yoshizaki, G. (2017). Application of dead end-knockout zebrafish as recipients of germ cell transplantation. Molecular Reproduction and Development, 84(10), 1100–1111. https://doi.org/10.1002/mrd.22870

Lightner, D.V. (1998). A handbook of shrimp pathology and diagnostic procedures for diseases of cultured penaeid shrimp. Baton Rouge, LA, USA, World Aquaculture Society. Lightner, D.V. & Redman, R.M. 1998. Shrimp disease and current diagnostic methods. Aquaculture, 164, 201–220. https://www.academia.edu/31148255/Shrimp_diseases_and_current_diagnostic_methods

Luo, Y. (Ed.). (2019). CRISPR gene editing: Methods and protocols. Humana Press. https://lib.ugent.be/en/catalog/ebk01:4930000000042208

Ma, J., Bruce, T. J., Jones, E. M., & Cain, K. D. (2019). A Review of Fish Vaccine Development Strategies: Conventional Methods and Modern Biotechnological Approaches. Microorganisms, 7(11), 569. https://doi.org/10.3390/microorganisms7110569

Mahendran, R., Jeyabaskar, S., Michael, D., Vincent Paul, A., & Sitharaman, G. (2016). Computer-aided vaccine designing approach against fish pathogens Edwardsiella tarda and Flavobacterium columnare using bioinformatics softwares. Drug Design, Development and Therapy, 1703. https://doi.org/10.2147/DDDT.S95691

Mair, G. C., Abucay, J. S., Abella, T. A., Beardmore, J. A., & Skibinski, D. (1997). Genetic manipulation of sex ratio for the large-scale production of all-male tilapia Oreochromis niloticus. Canadian Journal of Fisheries and Aquatic Sciences, 54(2), 396–404. https://doi.org/10.1139/f96-282

Malik, Y. P. S., Sagar, P., Dhama, K., & Singh, R. K. (2015). (eds) Current trends and future research challenges in vaccines and adjuvants. In: Souvenir, National Workshop Organized at Indian Veterinary Research Institute, Izatnagar 243122, Bareilly, Uttar Pradesh, India during 19–20 November, vol: pp. 1–130. https://www.researchgate.net/publication/285776754_Current_Trends_and_Future_Research_Challenges_in_Vaccines_and_Adjuvants

Melamed, P., Gong, Z., Fletcher, G., & Hew, C. L. (2002). The potential impact of modern biotechnology on fish aquaculture. Aquaculture, 204(3–4), 255–269. https://doi.org/10.1016/S0044-8486(01)00838-9

Meuwissen, T. H. E., Hayes, B. J., & Goddard, M. E. (2001). Prediction of Total Genetic Value Using Genome-Wide Dense Marker Maps. Genetics, 157(4), 1819–1829. https://doi.org/10.1093/genetics/157.4.1819

Morita, T., Morishima, K., Miwa, M., Kumakura, N., Kudo, S., Ichida, K., Mitsuboshi, T., Takeuchi, Y., & Yoshizaki, G. (2015). Functional Sperm of the Yellowtail (Seriola quinqueradiata) Were Produced in the Small-Bodied Surrogate, Jack Mackerel (Trachurus japonicus). Marine Biotechnology, 17(5), 644–654. https://doi.org/10.1007/s10126-015-9657-5

Morris, D. C., Morris, D. J., & Adams, A. (2002). Development of improved PCR to prevent false positives and false negatives in the detection of Tetracapsula bryosalmonae , the causative agent of proliferative kidney disease. Journal of Fish Diseases, 25(8), 483–490. https://doi.org/10.1046/j.1365-2761.2002.00398.x

Mousavi, S., Zahedinezhad, S., & Loh, J. Y. (2020). A review on insect meals in aquaculture: The immunomodulatory and physiological effects. International Aquatic Research, 12(2). https://doi.org/10.22034/iar(20).2020.1897402.1033

Nasopoulou, C., & Zabetakis, I. (2012). Benefits of fish oil replacement by plant originated oils in compounded fish feeds. A review. Lebensmittel-Wissenschaft und -Technologie, 47, 217–224. https://www.academia.edu/110089999/A_combined_in_vivo_and_in_vitro_approach_to_evaluate_the_influence_of_linseed_oil_or_sesame_oil_and_their_combination_on_innate_immune_competence_and_eicosanoid_metabolism_processes_in_common_carp_Cyprinus_carpio_

Noto, A., Dessi, A., Puddu, M., Mussap, M., & Fanos, V. (2014). Metabolomics technology and their application to the study of the viral infection. The Journal of Maternal-Fetal & Neonatal Medicine, 27(sup2), 53–57. https://doi.org/10.3109/14767058.2014.955963

Octavera, A., & Yoshizaki, G. (2019). Production of donor-derived offspring by allogeneic transplantation of spermatogonia in Chinese rosy bitterling†. Biology of Reproduction, 100(4), 1108–1117. https://doi.org/10.1093/biolre/ioy236

Okutsu, T., Shikina, S., Kanno, M., Takeuchi, Y., & Yoshizaki, G. (2007). Production of Trout Offspring from Triploid Salmon Parents. Science, 317(5844), 1517–1517. https://doi.org/10.1126/science.1145626

Okutsu, T., Takeuchi, Y., & Yoshizaki, G. (2008b) Spermatogonial transplantation in fish: production of trout offspring from salmon parents. In: Tsukamoto K et al (eds) Fisheries for global welfare and environment. TERRAPUB, Tokyo, pp 209–219. https://www.researchgate.net/publication/239784021_Spermatogonial_Transplantation_in_Fish_Production_of_Trout_Offspring_from_Salmon_Parents

Okutsu, T., Yano, A., Nagasawa, K., Shikina, S., Kobayashi, T., Takeuchi, Y., & Yoshizaki, G. (2006). Manipulation of Fish Germ Cell: Visualization, Cryopreservation and Transplantation. Journal of Reproduction and Development, 52(6), 685–693. https://doi.org/10.1262/jrd.18096

Overturf, K., LaPatra, S., & Powell, M. (2001). Real-time PCR for the detection and quantitative analysis of IHNV in salmonids. Journal of Fish Diseases, 24(6), 325–333. https://doi.org/10.1046/j.1365-2761.2001.00296.x

Pandian,T. J. & Sheela, S. G. (1995). Hormonal induction of sex reversal. Aquaculture,138, 1-22.http://www.petbh.com.br/guppy/wp-content/uploads/2017/12/Hormonal-induction-of-sex-reversal-in-fish.pdf

Pauly, D., & Zeller, D. (2017). Comments on FAOs State of World Fisheries and Aquaculture (SOFIA 2016). Marine Policy, 77, 176–181. https://doi.org/10.1016/j.marpol.2017.01.006

Poulson-Ellestad, K. L., Jones, C. M., Roy, J., Viant, M. R., Fernández, F. M., Kubanek, J., & Nunn, B. L. (2014). Metabolomics and proteomics reveal impacts of chemically mediated competition on marine plankton. Proceedings of the National Academy of Sciences, 111(24), 9009–9014. https://doi.org/10.1073/pnas.1402130111

Powell, J. L., & Loutit, M. W. (1994). Development of a DNA probe using differential hybridization to detect the fish pathogenVibrio anguillarum. Microbial Ecology, 28(3), 365–373. https://doi.org/10.1007/BF00662029

Pushpa, D., Yogendra, K. N., Gunnaiah, R., Kushalappa, A. C., & Murphy, A. (2014). Identification of Late Blight Resistance-Related Metabolites and Genes in Potato through Nontargeted Metabolomics. Plant Molecular Biology Reporter, 32(2), 584–595. https://doi.org/10.1007/s11105-013-0665-1

Puttinaowarat, S., Thompson, K.D. & Adams, A. (2000). Mycobacteriosis: Detection and identification of aquatic Mycobacterium species. Fish Vet. J., 5, 6–21. https://www.researchgate.net/publication/284544555_Mycobacteriosis_Detection_and_identification_of_aquatic_Mycobacterium_species

Pérez, O., Romeu, B., Cabrera, O., González, E., Batista-Duharte, A., Labrada, A., Pérez, R., Reyes, L. M., Ramírez, W., Sifontes, S., Fernández, N., & Lastre, M. (2013). Adjuvants are Key Factors for the Development of Future Vaccines: Lessons from the Finlay Adjuvant Platform. Frontiers in Immunology, 4. https://doi.org/10.3389/fimmu.2013.00407

Pérez-Sánchez, T., Balcázar, J. L., Merrifield, D. L., Carnevali, O., Gioacchini, G., De Blas, I., & Ruiz-Zarzuela, I. (2011). Expression of immune-related genes in rainbow trout (Oncorhynchus mykiss) induced by probiotic bacteria during Lactococcus garvieae infection. Fish & Shellfish Immunology, 31(2), 196–201. https://doi.org/10.1016/j.fsi.2011.05.005

Rao, Y., & Su, J. (2015). Insights into the Antiviral Immunity against Grass Carp (Ctenopharyngodon idella) Reovirus (GCRV) in Grass Carp. Journal of Immunology Research, 2015, 1–18. https://doi.org/10.1155/2015/670437

Rappuoli, R. (2000). Reverse vaccinology. Current Opinion in Microbiology, 3(5), 445–450. https://doi.org/10.1016/S1369-5274(00)00119-3

Rauta, P. R., Samanta, M., Dash, H. R., Nayak, B., & Das, S. (2014). Toll-like receptors (TLRs) in aquatic animals: Signaling pathways, expressions and immune responses. Immunology Letters, 158(1–2), 14–24. https://doi.org/10.1016/j.imlet.2013.11.013

Reid, G. K., & Jackson, T. (2014). Climate change sessions increasingly prominent at aquaculture meetings. Word Aquaculture, 45(3), 9-10. https://projects.upei.ca/climate/files/2012/07/2014-09-Climate-change-sessions-increaseingly-prominent-at-aquaculture-meetings.pdf

Richards, G. P., Watson, M. A., Needleman, D. S., Church, K. M., & Häse, C. C. (2015). Mortalities of Eastern and Pacific Oyster Larvae Caused by the Pathogens Vibrio coralliilyticus and Vibrio tubiashii. Applied and Environmental Microbiology, 81(1), 292–297. https://doi.org/10.1128/AEM.02930-14

Richards, R. G., Davidson, A. T., Meynecke, J.-O., Beattie, K., Hernaman, V., Lynam, T., & Van Putten, I. E. (2015). Effects and mitigations of ocean acidification on wild and aquaculture scallop and prawn fisheries in Queensland, Australia. Fisheries Research, 161, 42–56. https://doi.org/10.1016/j.fishres.2014.06.013

Ringø, E., Olsen, R. E., Jensen, I., Romero, J., & Lauzon, H. L. (2014). Application of vaccines and dietary supplements in aquaculture: Possibilities and challenges. Reviews in Fish Biology and Fisheries, 24(4), 1005–1032. https://doi.org/10.1007/s11160-014-9361-y

Roskoski, R., Lim, C. T., & Roskoski, L. M. (1975). Human brain and placental choline acetyltransferase: Purification and properties. Biochemistry, 14(23), 5105–5110. https://doi.org/10.1021/bi00694a013

Rumbos, C. I., Karapanagiotidis, I. T., Mente, E., & Athanassiou, C. G. (2019). The lesser mealworm Alphitobius diaperinus: A noxious pest or a promising nutrient source? Reviews in Aquaculture, 11(4), 1418–1437. https://doi.org/10.1111/raq.12300

Sahu, J. N., Karri, R. R., Zabed, H. M., Shams, S., & Qi, X. (2021). Current Perspectives and Future Prospects of Nano-Biotechnology in Wastewater Treatment. Separation & Purification Reviews, 50(2), 139–158. https://doi.org/10.1080/15422119.2019.1630430

Sakai, M. (1999). Current research status of fish immunostimulants. Aquaculture, 172(1–2), 63–92. https://doi.org/10.1016/S0044-8486(98)00436-0

Seibert, C. H., & Pinto, A. R. (2012). Challenges in shrimp aquaculture due to viral diseases: Distribution and biology of the five major penaeid viruses and interventions to avoid viral incidence and dispersion. Brazilian Journal of Microbiology, 43(3), 857–864. https://doi.org/10.1590/S1517-83822012000300002

Shao, Y., Li, C., Chen, X., Zhang, P., Li, Y., Li, T., & Jiang, J. (2015). Metabolomic responses of sea cucumber Apostichopus japonicus to thermal stresses. Aquaculture, 435, 390–397. https://doi.org/10.1016/j.aquaculture.2014.10.023

Sharif, M., Zafar, M. H., Aqib, A. I., Saeed, M., Farag, M. R., & Alagawany, M. (2021). Single cell protein: Sources, mechanism of production, nutritional value and its uses in aquaculture nutrition. Aquaculture, 531, 735885. https://doi.org/10.1016/j.aquaculture.2020.735885

Shen, Y., & Yue, G. (2019). Current status of research on aquaculture genetics and genomics-information from ISGA 2018. Aquaculture and Fisheries, 4(2), 43–47. https://doi.org/10.1016/j.aaf.2018.11.001

Shen, Y., Ma, K., & Yue, G. H. (2021). Status, challenges and trends of aquaculture in Singapore. Aquaculture, 533, 736210. https://doi.org/10.1016/j.aquaculture.2020.736210

Singh, R. K., Badasara, S. K., Dhama, K., & Malik, Y. P. S. (2015). Progress and Prospects in Vaccine Research. In National Workshop on "Current Trends and Future Research Challenges in Vaccines and Adjuvants"; Chapter, Organized at ICAR, Bareilly, Uttar Pradesh, India, 19–20 November 2015; Indian Veterinary Research Institute: Bareilly, Uttar Pradesh, India, 2015; pp. 1–19. https://scholar. google.com/scholar_lookup?title=Progress+and+ Prospects+in+Vaccine+Research&author=Singh, +R.K.&author=Badasara,+S.K.&author= Dhama,+K.&author=Malik,+Y.P.S. & publication_year=2015& pages=1%E2%80%9319

Starkey, W., Millar, R., Jenkins, M., Ireland, J., Muir, K., & Richards, R. (2004). Detection of piscine nodaviruses by real-time nucleic acid sequence based amplification (NASBA). Diseases of Aquatic Organisms, 59, 93–100. https://doi.org/10.3354/dao059093

Subasinghe, R. (2009). Disease control in aquaculture and the responsible use of veterinary drugs and vaccines: The issues, prospects and challenges. Options Mediterraneennes, A/no. 86. https://om.ciheam.org/om/pdf/a86/00801057.pdf

Subasinghe, R. P. & Bonad-Reamtaso, M. G. (2006). Biosecurity in Aquaculture: International Agreements and instruments, their compliance, prospects and challenges for Developing countries, pp.9-16. https://asthafoundation.in/img/2-Mohd.pdf

Tacon, A. G. J., & Metian, M. (2015). Feed Matters: Satisfying the Feed Demand of Aquaculture. Reviews in Fisheries Science & Aquaculture, 23(1), 1–10. https://doi.org/10.1080/23308249.2014.987209

Tang, J., Yuan, M., Wang, J., Li, Q., Huang, B., Wei, L., Liu, Y., Han, Y., Zhang, X., Wang, X., Zhang, M., & Wang, X. (2023). Identification and characterization of gonadotropin-releasing hormone (GnRH) in Zhikong scallop Chlamys farreri during gonadal development. Frontiers in Physiology, 14, 1180725. https://doi.org/10.3389/fphys.2023.1180725

The Aquaculture Genomics, Genetics and Breeding Workshop, Abdelrahman, H., ElHady, M., Alcivar-Warren, A., Allen, S., Al-Tobasei, R., Bao, L., Beck, B., Blackburn, H., Bosworth, B., Buchanan, J., Chappell, J., Daniels, W., Dong, S., Dunham, R., Durland, E., Elaswad, A., Gomez-Chiarri, M., Gosh, K., … Zhou, T. (2017). Aquaculture genomics, genetics and breeding in the United States: Current status, challenges, and priorities for future research. BMC Genomics, 18(1), 191, s12864-017-3557–1. https://doi.org/10.1186/s12864-017-3557-1

Thompson, K.D. & Adams, A. (2004). Current trends in immunotherapy and vaccine development for bacterial diseases of fish. In Molecular Aspects of Fish and Marine Biology: Current Trends in the Study of Bacterial and Viral Fish and Shrimp Diseases, 3: 313-362.

Turcios, A., & Papenbrock, J. (2014). Sustainable Treatment of Aquaculture Effluents—What Can We Learn from the Past for the Future? Sustainability, 6(2), 836–856. https://doi.org/10.3390/su6020836

Wang, L., Yue, F., Song, X., & Song, L. (2015). Maternal immune transfer in mollusc. Developmental & Comparative Immunology, 48(2), 354–359. https://doi.org/10.1016/j.dci.2014.05.010

Wang, Y.-S., & Shelomi, M. (2017). Review of Black Soldier Fly (Hermetia illucens) as Animal Feed and Human Food. Foods, 6(10), 91. https://doi.org/10.3390/foods6100091

Weber, G. M., & Lee, C.-S. (2014). Current and Future Assisted Reproductive Technologies for Fish Species. In G. C. Lamb & N. DiLorenzo (Eds.), Current and Future Reproductive Technologies and World Food Production (Vol. 752, pp. 33–76). Springer New York. https://doi.org/10.1007/978-1-4614-8887-3_3

Wei, L., Wang, Q., Ning, X., Mu, C., Wang, C., Cao, R., Wu, H., Cong, M., Li, F., Ji, C., & Zhao, J. (2015). Combined metabolome and proteome analysis of the mantle tissue from Pacific oyster Crassostrea gigas exposed to elevated pCO2. Comparative Biochemistry and Physiology Part D: Genomics and Proteomics, 13, 16–23. https://doi.org/10.1016/j.cbd.2014.12.001

Xu, P., David, L., Martínez, P., & Yue, G. H. (2020). Editorial: Genetic Dissection of Important Traits in Aquaculture: Genome-Scale Tools Development, Trait Localization and Regulatory Mechanism Exploration. Frontiers in Genetics, 11, 642. https://doi.org/10.3389/fgene.2020.00642

Yaqoob, A. A., Parveen, T., Umar, K., & Mohamad Ibrahim, M. N. (2020). Role of Nanomaterials in the Treatment of Wastewater: A Review. Water, 12(2), 495. https://doi.org/10.3390/w12020495

Yoshikawa, H., Takeuchi, Y., Ino, Y., Wang, J., Iwata, G., Kabeya, N., Yazawa, R., & Yoshizaki, G. (2017). Efficient production of donor-derived gametes from triploid recipients following intra-peritoneal germ cell transplantation into a marine teleost, Nibe croaker (Nibea mitsukurii). Aquaculture, 478, 35–47. https://doi.org/10.1016/j.aquaculture.2016.05.011

Yoshizaki, G., & Yazawa, R. (2019). Application of surrogate broodstock technology in aquaculture. Fisheries Science, 85(3), 429–437. https://doi.org/10.1007/s12562-019-01299-y

Yoshizaki, G., Ichikawa, M., Hayashi, M., Iwasaki, Y., Miwa, M., Shikina, S., & Okutsu, T. (2010). Sexual plasticity of ovarian germ cells in rainbow trout. Development, 137(8), 1227–1230. https://doi.org/10.1242/dev.044982

Yoshizaki, G., Okutsu, T., Ichikawa, M., Hayashi, M., & Takeuchi, Y. (2010b). Sexual plasticity of rainbow trout germ cells, Anim Reprod 7,187–196. http://www.cbra.org.br/pages/publicacoes/animalreproduction/issues/download/v7n3/pag187.pdf

Yue, G. H. (2014). Recent advances of genome mapping and marker-assisted selection in aquaculture. Fish and Fisheries, 15(3), 376–396. https://doi.org/10.1111/faf.12020

Zenger, K. R., Khatkar, M. S., Jones, D. B., Khalilisamani, N., Jerry, D. R., & Raadsma, H. W. (2019). Genomic Selection in Aquaculture: Application, Limitations and Opportunities With Special Reference to Marine Shrimp and Pearl Oysters. Frontiers in Genetics, 9, 693. https://doi.org/10.3389/fgene.2018.00693

Zhou, L., & Gui, J. (2018). Applications of Genetic Breeding Biotechnologies in Chinese Aquaculture. In J.-F. Gui, Q. Tang, Z. Li, J. Liu, & S. S. De Silva (Eds.), Aquaculture in China (1st ed., pp. 463–496). Wiley. https://doi.org/10.1002/9781119120759.ch6_1

Zhu, Z., Liu, G., He, L. & Chen, S. (1985). Novel gene transfer into fertilized eggs of goldfish (Carassius auratus L., 1758). Z. Angew. Ichthyol., 1: 31-34.https://patents.google.com/patent/CA2367129A1/en

Zokaeifar, H., Balcázar, J. L., Saad, C. R., Kamarudin, M. S., Sijam, K., Arshad, A., & Nejat, N. (2012). Effects of Bacillus subtilis on the growth performance, digestive enzymes, immune gene expression and disease resistance of white shrimp, Litopenaeus vannamei. Fish & Shellfish Immunology, 33(4), 683–689. https://doi.org/10.1016/j.fsi.2012.05.027

Zuma, A. (2022). Practical genetics and selective breeding in aquaculture. Conference: Arba Minch University PGS annual conference on Scientific improvements on livestock resources, April 2024. https://www.researchgate.net/publication/359649564_Practical_Genetics_and_Selective_Breeding_in_Aquaculture

2

Modern Biotechnological Approaches of Fish Based Biomaterials for Effective Medicinal Uses: An Update

Balaji Govindaswamy[1*] and Bharathi Ravikrishnan[2]

[1]*Department of Biotechnology, Bhupat and Jyoti Mehta School of Bioscience Indian Institute of Technology Madras. Chennai-600036, Tamil Nadu, India*

[2]*Department of Biotechnology, Guru Nanak College, Chennai-600042 Tamil Nadu India*

Abstract

The development of biomaterials, including implants, stents, customised drug delivery systems, tailored grafts, cell sheets, and other transplantable materials, has revolutionised health care and medicine in recent years. Biomaterials derived from natural biological sources, including extracellular proteins and other materials including collagen, gelatin, hyaluronic acid, peptides, and scales, are extensively utilised because to their notable advantages. Since collagen and gelatin have similar molecular structures and functions, gelatin is frequently employed in cell and tissue culture as a collagen substitute for biomaterial applications. The scales of fishes plays a vital role in wound healing applications which recently developed in healing the burn wounds, these scales are further analyzed in molecular level to study its peculiar applications towards human healthcare. Hyaluronic acid which is majorly produced in natural resources has potentially shown developments into hydrogels, scaffolds and drug delivery systems, especially from marine based HA has been developed into hydrogels to study the inner matrices and one of role player in tissue engineering. This chapter emphasis the challenges, current developments of the fish-based biomaterials offering a buildout in tissue engineering and healthcare technologies.

Keyword: *Collagen, Gelatin, Hyaluronic acid, Biomaterials and Tissue engineering.*

1. Introduction

Modern medicine relies heavily on biomaterials and tissue engineering, which are a broad category of materials designed to interact with biological systems in a therapeutic or diagnostic capacity. These materials, which are frequently polymers, natural or synthetic, are made to have particular physical, chemical, and biological characteristics in order to satisfy the demands of a range of medical applications (Holzapfel *et al.,* 2013; Ozdil & Aydin, 2014). These materials can be produced, modified and characterized in a number of ways, according to the particular needs of the application, including 3D printing, molding, and machining (Jandyal *et al.*, 2022). Biomaterials are extensively studied for their applications in tissue engineering, which comprises the recent developments into studying the disease models, micro-scale bone repairing, wound healing and other applications (Purnama *et al.*, 2010). Biomaterial degradation and biocompatibility must be taken into account when developing implants or medical devices since the material's characteristics are tunable according the external or internal parameters. In order to produce biomaterials for use in healthcare applications, cooperation between specialists in biology, medicine, and materials science is needed, the upcoming advances in bioprinting, artificial graft and usage of artificial intelligence might give us an insight towards development in disease regression (Holzapfel *et al.*, 2013; Khan *et al.*, 2015).

The majority of the protein found in the food supply worldwide comes from animal or grain sources, which together account for around half of the total. First of all, fish and crustaceans are significant and excellent providers of amino acids, a form of protein that is essential to nutrition and is only present in trace amounts in cereals and grains. It turns out that these matters for global nutrition and that some low-income, food-deficient nations really benefit from it. Furthermore, fisheries are significant local sources of food, commerce, and revenue in many coastal developing and developed countries. This has prompted researchers and biotechnologists to look at the identification, use, and extraction of these biomaterials in healthcare settings (Raghavendra *et al.*, 2015). The varied fish-derived biomaterials have a variety of uses, including as in tissue engineering, medication delivery, and wound healing, which has opened up new avenues for medical science to make scientific discoveries (Bhat & Kumar, 2013).

These biomaterials, which are derived from fish sources such scales, skin, bones, and gelatin, fig.1 have remarkable biocompatibility and, in some situations, characteristics that are similar to those of human tissues. The fish biowaste mainly composed of collagen type-I, hydroxyapatite, hyaluronic acid, keratin, polysaccharides (such as alginate, cellulose, chitin, chitosan chondroitin sulfate and starch), terpenes (such as natural rubber) and lipids are been used extensively due to their biocompatibility (Di *et al.*, 2022). These biomaterials have been used in tissue engineering, wound healing, and cosmetic formulations (Rezvani Ghomi *et al.*, 2021). Their favourable characteristics, biodegradability, and sustainable sourcing make them essential contributors to the biomaterials landscape, highlighting the significance of using nature's resources to advance medical science and providing creative solutions for difficult healthcare problems (Van Den Bosch & Ode Sang, 2017).

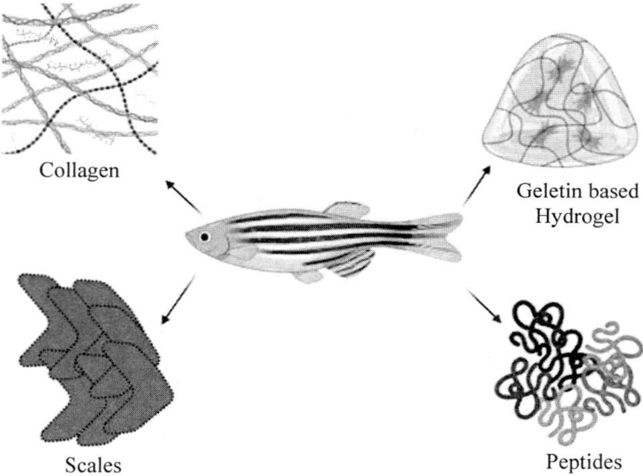

Fig. 1. Diverse materials of marine fish

This chapter aims to provide a current review of the state-of-the-art biotechnological techniques utilized in fish-based biomaterials and medicinal applications. These biomaterials' unique properties have the ability to address unmet medical needs and inspire innovative thinking in the quickly evolving field of healthcare technology.

2. Fish Based Materials And Their Applications

2.1 Collagen

Collagen is a necessary protein found in living things that creates the structure of bones, cartilage, cartilage capsules, joints, tendons, skin, hair, and nails. Collagen is a part of the connective tissue layer of the extracellular matrix. Collagen comes in several forms, depending on the function and geographical area (Kanta, 2015). Marine animals, including whales, and invertebrates, including sponges, jellyfish, starfish, squids, and sea urchins, have all been used to harvest collagen. Within the marine vertebrate group, fish have become the primary source of marine collagen (Felician *et al.*, 2018). This is because fish are the most eaten marine organisms worldwide, and the majority of the marine waste from the fish processing industry is composed of collagen-rich fish parts. Sponge and jellyfish are the most often utilized marine invertebrate sources. In a nutshell, enzymatic treatments or chemical solubilization using acids, bases, or neutral solutions are often used in extraction methods. The most popular technique is acetic acid solubilization, which is followed by sodium chloride (NaCl) salting out. Supercritical fluid (SCF), a cutting-edge green technology, has also been investigated as an extraction method to encourage the solubilization of collagen from different marine sponge species (Y. Wu *et al.*, 2022). The construction of scaffolds that mimic the intricate structural hierarchy and mechanical integrity of original tissue has been made possible by advancements in fabrication technology (Collins *et al.*, 2021). Marine collagen scaffolds are often created via electrospinning, hydrogelation, 3D bioprinting, and decellularization. Because of its three-dimensional (3D) water-swollen network, which mimics the characteristics of natural extracellular matrix (ECM), hydrogels have garnered a lot of interest in the field of tissue engineering (Fan & Wang, 2017). It has been demonstrated that the manufacturing parameters collagen source, collagen concentration, pH, ionic strength, and temperature affect the characteristics of the collagen hydrogel. Marine collagen hydrogels are less stable than hydrogels made from mammalian collagen because there are less amino acids available for intermolecular hydrogen bonding, despite the fact that marine collagen may spontaneously self-assemble into fibrillar structures. As a result, crosslinking has been extensively employed to enhance the marine collagen hydrogels' mechanical characteristics and degradation profile (Bhattacharjee & Ahearne, 2021).

In addition to its ease of use, electrospinning is prized for its capacity to create porous scaffolds with micro- to nanoscale threads that replicate the porosity structure of natural extracellular matrix (ECM) and high surface area (Ashammakhi *et al.*, 2008). It's critical to select a carrier solvent with the right viscosity, volatility, and conductivity to produce homogenous, smooth fibers. Fluoro-alcohols, such hexafluoro-isopropanol (HFIP), are frequently employed for the electrospinning of marine collagen because of their volatile nature and capacity to breakdown collagen into a viscous, conductive solution that is ideal for electrospinning (Liu *et al.*, 2022). Mechanically stronger polymers, such polycaprolactone (PCL), can be added to the collagen electrospinning solution since HFIP can dissolve a broad variety of polymers (Ingavle & Leach, 2014). This will result in composite fibers that are stronger than fibers electrospun with seawater.

Bone fracture healing and repair are examples of postnatal regenerative processes that may be thought of as tissue regeneration processes that mimic many aspects of bone growth. As a result, bone regeneration is a multifaceted physiological process that includes both bone resorption and bone creation. Avascular necrosis, atrophic non-unions, and osteoporosis are examples of complex clinical conditions where a significant amount of bone regeneration is necessary (Dimitriou *et al.*, 2011). Other conditions include skeletal reconstruction of large bones that have been damaged or defective by trauma, infection, tumor resection, or skeletal abnormalities.

2.3 Scales

Fish scales, which are solid plates that are affixed to the fish's skin, have a certain combination of strength, flexibility, and toughness that is necessary to protect the fish without impairing its mobility. Type I collagen, calcium-deficient hydroxyapatite inorganic phase, lecithin organic phase, and a few other trace elements make up the majority of the fish scales (Kattimani *et al.*, 2016). Fish scales have similar traits in their basic structure, but they can vary in form and ornamentation depending on a variety of parameters, including fish species, behaviors, ages, and sizes. The rear portion of the scales covers the anterior portion, which is buried in the skin. Radial grooves split the whole fish scale into distinct scalloped regions. Divergent opinions exist on the microstructure of fish scales in cross section. According to some studies, the layers of collagen that are pliable and mineralized make fish scales. The inner collagen layer is organized in a thin-layer plywood structure with a low degree of mineralization, whereas the outside bone layer is formed in an ordered, continuous, rough structure (Jia *et al.*, 2022).

As its name implies, HA is the hydroxylated form of the phosphate minerals, which are characteristic of the apatite family. These bioceramics may be intentionally synthesized using a variety of procedures, such as precipitation, hydrothermal, multiple emulsion, biomimetic deposition, and electrodeposition methods (Babaie & Bhaduri, 2018; Panda *et al.*, 2021). They crystallize into the hexagonal system. Additionally, HA powders and coatings have been effectively created using the sol gel method, which combines various calcium and phosphorus precursor mixtures to create a high molecular purity of HA. Since hydroxyapatite is biocompatible, bioactive, bone conductive, and non-toxic to living things, it has garnered a lot of interest in the field of bone tissue engineering (Kumar *et al.*, 2020). By using chemical solution coprecipitation, the researchers were able to evaluate the physicochemical, bioactivities, and biological characteristics of hydroxyapatite obtained from fish scale with synthetic hydroxyapatite.

In order to fend against predators throughout their natural development, several species have developed body armor. A common example is the teleost, which has "plate-like" elastic scales made of collagen that are translucent, flexible, lightweight, and resistant to penetration. Fish skin's surface is covered with scales that have evolved over millions of years into an incredible array of different hierarchical structures. The highly organized microstructure of those fish scales shows a superb blend of mechanical strength and toughness. Because natural or synthetic scales resemble human bones and teeth so closely, they are frequently employed as materials for bone tissue engineering (Sherman *et al.*, 2015 Ghazlan *et al.*, 2021; Rawat *et al.*, 2021;).

The fish scales and skin which are the part of the integumentary system are also good source of pigments, which can be used in cosmetic, food industries, due to its biocompatibility (Salvatore *et al.,* 2020). The pigments are stored in the chromophores. Eight types of fish chromophores have been detected so far and they are melanophore (black), xanthophore (yellow to orange), erythrophore (orange to red), iridophore (iridescent, blue or silver), leucophore (white, reflective), cyanophore (blue), erythro-iridophore (reddish violet) and erythro-cyanophore (bluish and reddish) (Wenyu *et al.*, 2022). The fish skin also has mineral deposits which can be extracted and widely used in the cosmetics industries. Mineral such as garnet, mica, quartz, amphibole are present as microstructures in the fish scales (Bruno *et al.*, 2023).

2.3 Gelatin

Fascia that has been partly hydrolyzed is called gelatin. Studies have shown that they may be extracted from marine sources including fish and sponges, even though their primary sources are the skins of cattle and pigs (Pratt, 2021). Gelatin's chemical makeup varies depending on the source, however it is more likely to contain hydrophobic amino acids such as glycine (Gly), hydroxyproline (Hyp), and proline (Pro) (Z. Wu *et al.*, 2019). Fascia may be heated in an acidic or alkaline solution to turn it into soluble gelatin. The breaking of many intra- and intermolecular covalent crosslinks found in collagen is the cause of its thermal solubilization in the presence of acid or alkali (Rýglová *et al.*, 2017). Additionally, hydrolysis occurs to a few amide bonds in the collagen molecules' basic chains. Fish gelatin can also be utilised in applications that don't need gelling or a high Bloom value because of its other qualities, which include texturization and syneresis avoidance (Venugopal, 2021). Fish gelatin from cold water can be utilised in chilled or frozen goods that are immediately eaten after being taken out of the refrigerator or defrosted. Fish gelatin has additional potential uses because to its low gelling temperature (Wasswa *et al.*, 2007). Low melting point gelatins may also be utilised in dry goods (such micro-encapsulation). In fact, fish gelatin is mostly used to microencapsulate vitamins and other medicinal additives like azoxanthine (Karim & Bhat, 2009). A prospective source of gelatin for creating other natural-based polymers, such as the foundational ingredient for hydrogel fabrications, is fish gelatin. Fish gelatin-based hydrogels are made either from pure fish gelatin or by changing a functional group in fish gelatin to create fish gelatin hybrid hydrogels (Huang *et al.*, 2019). Fish gelatin composite hydrogels, on the other hand, are hydrogels made of fish gelatin combined with one or more other polymers or materials. Fish gelatin is being modified in order to create a network precursor that can be cross-linked and polymerized more readily. In order to create fish gelatinmethacrylate (fish GelMA) hydrogels, methacrylic anhydrate is often combined with fish gelatin in co-polymer or composite materials (Salahuddin *et al.*, 2021). This co-polymer is most likely the result of several investigations on the hybridization of methacrylic acid with conventional mammalian gelatin to provide the appropriate mechanical or physical characteristics of hydrogels. For 3D printing applications, even mammalian gelatin methacrylic (GelMA) has been commercialised as one of the finest bioinks (Hölzl *et al.*, 2016).

2.4 Peptides

Numerous marine species encompass peptides, which are significant bioactive natural compounds about which much research has been done. The sources and functions of bioactive peptides in marine animals are mostly unknown. Their potent bioactivities, which include anticancer, antidiabetic, neroprotective, and cardioprotective properties, are unrelated to their in situ functions (Hao *et al.*, 2020). The identification of these bioregulatory functions and the clarification of the marine peptides' methods of action support the use of the peptides as possible medications for the treatment of diabetes, hypertension, and cancer (Cicero *et al.*, 2017). Marine peptides' wide range of bioactivity has great promise for medical and nutraceutical benefits, drawing interest from the pharmaceutical and nutraceutical industries in the hopes that they might be applied to the prevention or treatment of a number of illnesses (Nelson *et al.*, 2010). The creation of appropriate scaffold materials that support structure and encourage cell growth is one of the main issues facing tissue engineering (C. Liu *et al.*, 2007). Scaffold materials can benefit from the addition of marine peptides to improve their biocompatibility and bioactivity, which include their capacity to replicate the extracellular matrix and encourage cell attachment (B. Li *et al.*, 2020). Though there are currently more natural products on the market, very few of these substances are commercially available (Li & Vederas, 2009). A small number of discovered peptides from marine creatures are undergoing preclinical testing, and a few of them have advanced to various stages of clinical testing to demonstrate their potential as anticancer medications.

Conclusion

This exploration into marine based resources for alternate usage towards healthcare underscores the promising strides made in the field of medicine. The development of innovative drug delivery systems using fish-derived biomaterials offers unprecedented opportunities for precision medicine and targeted therapeutic interventions. However, challenges such as sustainability, ethical considerations and scalability need carefully considered for these processes. In essence, the chapter defines the usage of materials towards targeted problem where the convergence of biology, technology, and medicine opens new frontiers for healing, regeneration and improved healthcare outcomes.

References

Ashammakhi, N., Ndreu, A., Nikkola, L., Wimpenny, I., & Yang, Y. (2008). Advancing tissue engineering by using electrospun nanofibers. Regenerative Medicine, 3(4), 547–574. https://doi.org/10.2217/17460751.3.4.547

Babaie, E., & Bhaduri, S. B. (2018). Fabrication Aspects of Porous Biomaterials in Orthopedic Applications: A Review. ACS Biomaterials Science & Engineering, 4(1), 1–39. https://doi.org/10.1021/acsbiomaterials.7b00615

Bhat, S., & Kumar, A. (2013). Biomaterials and bioengineering tomorrow's healthcare. Biomatter, 3(3), e24717. https://doi.org/10.4161/biom.24717

Bhattacharjee, P., & Ahearne, M. (2021). Significance of Crosslinking Approaches in the Development of Next Generation Hydrogels for Corneal Tissue Engineering. Pharmaceutics, 13(3), 319. https://doi.org/10.3390/pharmaceutics13030319

Bruno Riberiro., V, Christopher Kirkland., L, Melanie Finch., A, Frederico Faleiros., M, Steven Reddy., M, William Rickard., D.A. and Michael Hartnady, I. H. (2023). Microstructures, geochemistry and geochronology of mica fish: review and advances. Journal of structural geology, Vol.175. 104947. https://doi.org/10.1016 /j.jsg. 2023.104947

Cicero, A. F. G., Fogacci, F., & Colletti, A. (2017). Potential role of bioactive peptides in prevention and treatment of chronic diseases: A narrative review. British Journal of Pharmacology, 174(11), 1378–1394. https://doi.org/10.1111/bph.13608

Collins, M. N., Ren, G., Young, K., Pina, S., Reis, R. L., & Oliveira, J. M. (2021). Scaffold Fabrication Technologies and Structure/Function Properties in Bone Tissue Engineering. Advanced Functional Materials, 31(21), 2010609. https://doi.org/10.1002/adfm.202010609

Dimitriou, R., Jones, E., McGonagle, D., & Giannoudis, P. V. (2011). Bone regeneration: Current concepts and future directions. BMC Medicine, 9(1), 66. https://doi.org/10.1186/1741-7015-9-66

Din Qin, Shichao Bi, Xinguo You, Mengyang Wang, Xin Cong, Congshan Yuan, Miao Yu, Xi aojie Cheng, Xi-Guang Chen. Development and application of fish scale wastes as versatile natural biomaterials. Chemical Engineering Journal, 428, 131102. https://doi.org/10.1016/j.cej.2021.131102

Fan, C., & Wang, D.-A. (2017). Macroporous Hydrogel Scaffolds for Three-Dimensional Cell Culture and Tissue Engineering. Tissue Engineering Part B: Reviews, 23(5), 451–461. https://doi.org/10.1089/ten.teb.2016.0465

Felician, F. F., Xia, C., Qi, W., & Xu, H. (2018). Collagen from Marine Biological Sources and Medical Applications. Chemistry & Biodiversity, 15(5), e1700557. https://doi.org/10.1002/cbdv.201700557

Ghazlan, A., Ngo, T., Tan, P., Xie, Y. M., Tran, P., & Donough, M. (2021). Inspiration from Nature's body armours – A review of biological and bioinspired composites. Composites Part B: Engineering, 205, 108513. https://doi.org/10.1016/j.compositesb.2020.108513

Hao, M., Lv, M., & Xu, H. (2020). Andrographolide: Synthetic Methods and Biological Activities. Mini-Reviews in Medicinal Chemistry, 20(16), 1633–1652. https://doi.org/10.2174/1389557520666200429100326

Holzapfel, B. M., Reichert, J. C., Schantz, J.-T., Gbureck, U., Rackwitz, L., Nöth, U., Jakob, F., Rudert, M., Groll, J., & Hutmacher, D. W. (2013). How smart do biomaterials need to be? A translational science and clinical point of view. Advanced Drug Delivery Reviews, 65(4), 581–603. https://doi.org/10.1016/j.addr.2012.07.009

Huang, T., Tu, Z., Shangguan, X., Sha, X., Wang, H., Zhang, L., & Bansal, N. (2019). Fish gelatin modifications: A comprehensive review. Trends in Food Science & Technology, 86, 260–269. https://doi.org/10.1016/j.tifs.2019.02.048

Hölzl, K., Lin, S., Tytgat, L., Van Vlierberghe, S., Gu, L., & Ovsianikov, A. (2016). Bioink properties before, during and after 3D bioprinting. Biofabrication, 8(3), 032002. https://doi.org/10.1088/1758-5090/8/3/032002

Ingavle, G. C., & Leach, J. K. (2014). Advancements in Electrospinning of Polymeric Nanofibrous Scaffolds for Tissue Engineering. Tissue Engineering Part B: Reviews, 20(4), 277–293. https://doi.org/10.1089/ten.teb.2013.0276

Jandyal, A., Chaturvedi, I., Wazir, I., Raina, A., & Ul Haq, M. I. (2022). 3D printing – A review of processes, materials and applications in industry 4.0. Sustainable Operations and Computers, 3, 33–42. https://doi.org/10.1016/j.susoc.2021.09.004

Jia, Z., Deng, Z., & Li, L. (2022). Biomineralized Materials as Model Systems for Structural Composites: 3D Architecture. Advanced Materials, 34(20), 2106259. https://doi.org/10.1002/adma.202106259

Kanta, J. (2015). Collagen matrix as a tool in studying fibroblastic cell behavior. Cell Adhesion & Migration, 9(4), 308–316. https://doi.org/10.1080/19336918.2015.1005469

Karim, A. A., & Bhat, R. (2009). Fish gelatin: Properties, challenges, and prospects as an alternative to mammalian gelatins. Food Hydrocolloids, 23(3), 563–576. https://doi.org/10.1016/j.foodhyd.2008.07.002

Kattimani, V. S., Kondaka, S., & Lingamaneni, K. P. (2016). Hydroxyapatite—Past, Present, and Future in Bone Regeneration. Bone and Tissue Regeneration Insights, 7, BTRI.S36138. https://doi.org/10.4137/BTRI.S36138

Khan, F., Tanaka, M., & Ahmad, S. R. (2015). Fabrication of polymeric biomaterials: A strategy for tissue engineering and medical devices. Journal of Materials Chemistry B, 3(42), 8224–8249. https://doi.org/10.1039/C5TB01370D

Kumar, P., Saini, M., Dehiya, B. S., Sindhu, A., Kumar, V., Kumar, R., Lamberti, L., Pruncu, C. I., & Thakur, R. (2020). Comprehensive Survey on Nanobiomaterials for Bone Tissue Engineering Applications. Nanomaterials, 10(10), 2019. https://doi.org/10.3390/nano10102019

Li, B., Elango, J., & Wu, W. (2020). Recent Advancement of Molecular Structure and Biomaterial Function of Chitosan from Marine Organisms for Pharmaceutical and Nutraceutical Application. Applied Sciences, 10(14), 4719. https://doi.org/10.3390/app10144719

Li, J. W.-H., & Vederas, J. C. (2009). Drug Discovery and Natural Products: End of an Era or an Endless Frontier? Science, 325(5937), 161–165. https://doi.org/10.1126/science.1168243

Liu, C., Xia, Z., & Czernuszka, J. T. (2007). Design and Development of Three-Dimensional Scaffolds for Tissue Engineering. Chemical Engineering Research and Design, 85(7), 1051–1064. https://doi.org/10.1205/cherd06196

Liu, S., Lau, C.-S., Liang, K., Wen, F., & Teoh, S. H. (2022). Marine collagen scaffolds in tissue engineering. Current Opinion in Biotechnology, 74, 92–103. https://doi.org/10.1016/j.copbio.2021.10.011

Nelson, J. E., Cox, C. E., Hope, A. A., & Carson, S. S. (2010). Chronic Critical Illness. American Journal of Respiratory and Critical Care Medicine, 182(4), 446–454. https://doi.org/10.1164/rccm.201002-0210CI

Ozdil, D., & Aydin, H. M. (2014). Polymers for medical and tissue engineering applications: Polymers for medical and tissue engineering applications. Journal of Chemical Technology & Biotechnology, 89(12), 1793–1810. https://doi.org/10.1002/jctb.4505

Panda, S., Biswas, C. K., & Paul, S. (2021). A comprehensive review on the preparation and application of calcium hydroxyapatite: A special focus on atomic doping methods for bone tissue engineering. Ceramics International, 47(20), 28122–28144. https://doi.org/10.1016/j.ceramint.2021.07.100

Pratt, R. L. (2021). Hyaluronan and the Fascial Frontier. International Journal of Molecular Sciences, 22(13), 6845. https://doi.org/10.3390/ijms22136845

Purnama, A., Hermawan, H., Couet, J., & Mantovani, D. (2010). Assessing the biocompatibility of degradable metallic materials: State-of-the-art and focus on the potential of genetic regulation. Acta Biomaterialia, 6(5), 1800–1807. https://doi.org/10.1016/j.actbio.2010.02.027

Raghavendra, G. M., Varaprasad, K., & Jayaramudu, T. (2015). Biomaterials. In Nanotechnology Applications for Tissue Engineering (pp. 21–44). Elsevier. https://doi.org/10.1016/B978-0-323-32889-0.00002-9

Rawat, P., Zhu, D., Rahman, M. Z., & Barthelat, F. (2021). Structural and mechanical properties of fish scales for the bio-inspired design of flexible body armors: A review. Acta Biomaterialia, 121, 41–67. https://doi.org/10.1016/j.actbio.2020.12.003

Rezvani Ghomi, E., Nourbakhsh, N., Akbari Kenari, M., Zare, M., & Ramakrishna, S. (2021). Collagen-based biomaterials for biomedical applications. Journal of Biomedical Materials Research Part B: Applied Biomaterials, 109(12), 1986–1999. https://doi.org/10.1002/jbm.b.34881

Rýglová, Š., Braun, M., & Suchý, T. (2017). Collagen and Its Modifications-Crucial Aspects with Concern to Its Processing and Analysis. Macromolecular Materials and Engineering, 302(6), 1600460. https://doi.org/10.1002/mame.201600460

Salahuddin, B., Wang, S., Sangian, D., Aziz, S., & Gu, Q. (2021). Hybrid Gelatin Hydrogels in Nanomedicine Applications. ACS Applied Bio Materials, 4(4), 2886–2906. https://doi.org/10.1021/acsabm.0c01630

Salvatore, L., Gallo, N., Natali, M. L., Campa, L., Lunetti, P., Madaghiele, M., Blasi, F. S., Corallo, A., Capobianco, L., & Sannino, A. (2020). Marine collagen and its derivatives: Versatile and sustainable bio-resources for healthcare. Materials Science and Engineering: C, 113, 110963. https://doi.org/10.1016/j.msec.2020.110963

Sherman, V. R., Yang, W., & Meyers, M. A. (2015). The materials science of collagen. Journal of the Mechanical Behavior of Biomedical Materials, 52, 22–50. https://doi.org/10.1016/j.jmbbm.2015.05.023

Van Den Bosch, M., & Ode Sang, Å. (2017). Urban natural environments as nature-based solutions for improved public health – A systematic review of reviews. Environmental Research, 158, 373–384. https://doi.org/10.1016/j.envres.2017.05.040

Venugopal, V. (2021). Valorization of Seafood Processing Discards: Bioconversion and Bio-Refinery Approaches. Frontiers in Sustainable Food Systems, 5, 611835. https://doi.org/10.3389/fsufs.2021.611835

Wasswa, J., Tang, J., & Gu, X. (2007). Utilization of Fish Processing By-Products in the Gelatin Industry. Food Reviews International, 23(2), 159–174. https://doi.org/10.1080/87559120701225029

Wenyu Fang, Junrou Huang, Shizhu Li, Jianguo Lu. (2022). Identification of pigment genes (melanin, carotenoid and pteridine) associated with skin colour variant in red tilapia using trandcriptome analysis. Aquaculture, Vol.547 (30), https://doi.org/10.1016/j.aquaculture.2021.737429.

Wu, Y., Wu, Z.-M., Zhang, S.-S., Liu, L.-Y., Sun, F., Jiao, W.-H., Wang, S.-P., & Lin, H.-W. (2022). Axinellasins A–D, Immunosuppressive Cycloheptapeptide Diastereomers, Discovered via a Precursor Ion Scanning–Supercritical Fluid Chromatography Strategy from the Marine Sponge Axinella species. Organic Letters, 24(3), 934–938. https://doi.org/10.1021/acs.orglett.1c04309

Wu, Z., Hou, Y., Dai, Z., Hu, C.-A. A., & Wu, G. (2019). Metabolism, Nutrition, and Redox Signaling of Hydroxyproline. Antioxidants & Redox Signaling, 30(4), 674–682. https://doi.org/10.1089/ars.2017.7338

3

"Blue Biotech Revolution Next-Gene Sequencing in Fisheries and Aquaculture"

***Arya Singh*[1], *Gowhar Iqbal*[1*], *Vivek Chauhan*[2], *Mahvish Mehdi*[3], and *Lukram Sushil*[1]**

[1]*ICAR-Central Institutes of Fisheries Education, Mumbai-400061, Maharashtra, India*

[2]*College of Fisheries Science, CCS HAU, Hisar-12500, Haryana India*

[3]*Era University, Tondan Marg, Lucknow-226003, Uttar Pradesh, India*

Abstract

Recent technological advances in nucleic acid sequencing, known as next-generation sequencing (NGS), have revolutionized the field of genomics. Next-generation sequencing (NGS) technologies have transformed not only genomics but also many other fields because of their rapid progress and falling costs. In the realm of fisheries and aquaculture, NGS has provided a paradigm shift from gene to genome-wide research, leading to an astounding number of scientific discoveries. NGS has been used by fish biologists to identify new genetic markers for population and traceability research. NGS technology has also been applied in ecotoxicological applications and genome-wide characterization to identify genomic areas linked to commercially significant features. These new technologies have made significant contributions to the field of fish pathogen research. NGS is widely used to track aquatic viruses, examine the virulence and evolutionary history of diseases, and identify new etiological agents that cause fatalities. This chapter provides a brief description of these technologies and examples of how they are used in aquaculture.

Keywords: *next-generation sequencing (NGS), genomics, aquaculture, ecotoxicological*

1. Introduction

With the advancement of science and technology, new technologies have been developed to advance molecular work to a great extent. Thousands of DNA molecules can be simultaneously sequenced at one time using next-generation sequencing (NGS). Next-generation sequencing (NGS) technologies act as revolutionary tools for various implementations and can produce hundreds of gigabases of sequence data within a single experiment. A wide variety of modern sequencing technologies go by the name of "high-throughput sequencing." In recent years, this technology has advanced as a de facto tool for the study of genomics and epigenomics. Nucleotides can be located using this technique anywhere in the genome or in groups of specific DNA or RNA regions. An NGS plot is a useful tool for bridging the gap between enormous datasets and genomic information in this era of massive sequencing data. Resequencing of genomic DNA will undoubtedly be the primary application of NGS in clinical settings. Whole genome sequencing (WGS) offers an intricate representation of single nucleotide variations (SNV), insertions and deletions (indels), complex structural alterations, and changes in copy number, all of which are attainable through a single analysis. Extensive genetic information can be obtained and used after complete genome sequencing, which reveals a large amount of genetic data. However, a significant portion contains either novel information or data whose clinical significance has not yet been established. WGS is the most comprehensive genetic analysis of an individual or cancer genome.

Because of the large need for low-cost sequencing, high-throughput sequencing has been created, producing hundreds or even millions of sequences at once. They are made to cut DNA sequencing costs even lower than those of traditional dye-terminator methods. Because we can now sequence DNA and RNA far more quickly and cheaply than we could with Sanger sequencing, these new technologies have transformed the study of genomics and molecular biology. NGS approaches, therefore, consider technologies that have emerged subsequent to the Sanger method. Although Sanger sequencing is gradually being replaced by NGS in molecular diagnostics, the two methods have a similar history that goes back millions of years; both repurpose the DNA replication machinery that copies DNA during every cell division. These innovations include sequencing-by-synthesis, sequencing-by-ligation, and ion-semiconductor sequencing.

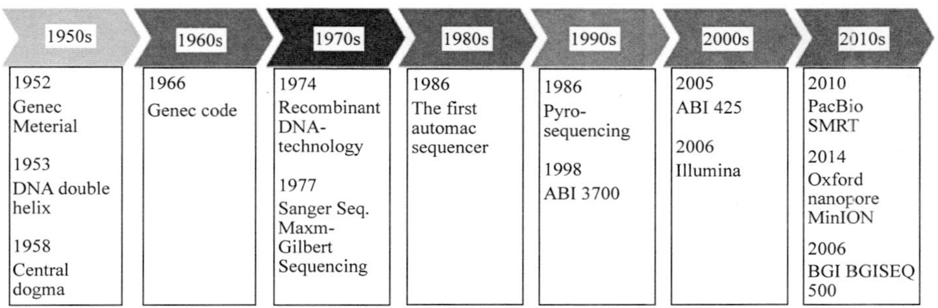

Fig. 1. Evolution of the sequencing methods in chronological order (Pervez *et al.*, 2022)

By allowing billions of sequencing operations to occur simultaneously on a single surface (glass or beads), next-generation sequencing (NGS) dramatically increases throughput while lowering costs compared to first-generation sequencing.

1.1 Illumina: One of the top producers of several sequencing tools, is Illumina Inc. (Illumina, 2022). It was founded in San Diego, California in 1998. Currently, it offers several sequencing platforms divided into two categories: Production Scale Sequencers (PSS) and Benchtop Sequencers (BTS). All BTS offer assistance for: (1) whole genome sequencing (WGS) for viruses and bacteria, (2) target gene sequencing (TGS), (3) target gene expression profiling (TGEP), (4) miRNA and sRNA analysis profiling, and (5) 16S metagenomic sequencing (MS) (except for iSeq100).

1.2 Ion Torrent: Ion Torrent was introduced in 2011 (Rothberg *et al.*, 2011). Ion Torrent is a sequencing by synthesis (SBS) method that generates nucleotide sequences by measuring pH. The sequencing reads produced by Ion Torrent varied in length. Sequencing from either end of a fragment cannot be performed using Ion Torrent sequencing equipment (Lahens *et al.*, 2017). The GeneXus system, Ion GeneStudio S5 system, PGM Dx system, and Ion Chef system are the four Ion Torrent instruments.

1.3 GenapSys: Founded in 2010, GenapSys is a firm affiliated with the Stanford Genome Technology Center. By integrating the thermal detection of nucleotide mutations, the GenapSys Sequencer improved the sequencing by synthesis (SBS) method (Niedringhaus *et al.*, 2011). It is inexpensive, lightweight (less than ten pounds), simple to operate, and suitable even for those new to the field of genomics.

1.4 QIAGEN: For the creation of NGS data, QIAGEN offers a GeneReader. The nucleotides were located using fluorescent signal templates that matched and were clonally amplified using the Gene-Read QIAcube. The GeneReader system facilitates all phases of sample processing and sequencing, including

DNA extraction, library preparation, sequencing, bioinformatic analysis, clinical implications, and proof.

1.5 Roche 45: The Roche GS-FLX 454 Genome Sequencer was the first commercial system released as a 454 Sequencer in 2004 (Mostafa *et al.*, 2015). James D. Watson's genome was the second person whose entire genome was sequenced using this method. The average read length and accuracy of the improved 454 GS FLX Titanium technology, released in 2008, increased to 700 bp and 99.997%, respectively. In less than a day, this platform increased the data output by 0.7 gigabyte per run.

Table 1. Comparative analysis of many high-performance sequencing machines (Pervez *et al.*, 2022).

S.No.	Manufacturer	Read length	Data output	Max. run time (hours)	Chemistry
1.	(NovaSeq 6000) from Illumina	300 PE	6Tb (6000 Gb)	44	Sequencing by synthesis
2.	(Ion GeneStudio S5 Prime) from Ion Torrent	600 SE	50 Gb	12	Sequencing by synthesis
3.	(16 chips) in GenapSys	150 SE	2 Gb	24	Sequencing by synthesis
4.	QIAGEN (GeneReader)	100 SE	-	-	Sequencing by synthesis
5.	Nanopore (PromethION)	4Mb	14Tb (14000 Gb)	72	Real-time sequencing

2. Types of NGS

2.1 Massively parallel signature sequencing (MPSS) by Lynx therapeutics

This "next-generation" sequencing technology was regarded as the first. The MPSS was developed in the 1990s at Lynx Therapeutics, a company founded in 1992 by Sydney Brenner and Sam Eletr. This sequencing process has high throughput. It determines the exact expression level of each transcript in the sample, and makes almost all of them visible when used to generate an expression profile. Because the MPSS bead-based approach uses a complex strategy of adapter ligation followed by adapter decoding and reading the sequence in increments of four nucleotides, it is susceptible to sequence-specific bias or loss of specific sequences. Nonetheless, the core features of the MPSS output, comprising several short DNA sequences, are indicative of subsequent "next-gen" data formats.

Massive parallel signature sequencing (MPSS) is a technique developed to sequence mRNA transcripts and measure gene expression based on individual mRNA levels in the cell. MPSS finds transcripts on individual microbeads using a complementary DNA signature sequence. The microbeads were analyzed in an array format using a flow cell. The bases of the mRNA were carefully scanned and removed by hybridization to a fluorescently labeled coder. Rounds of identification and removal were performed one after the other until the process was completed. An array of sequences ranging in length from 17 to 20 bp was obtained. Thousands of transcripts were simultaneously sequenced. The transcript count, a measure of expression level, is based on the number of transcripts per million molecules. MPSS does not require identification and characterization of genes prior to the commencement of the investigation. MPSS is sensitive to only a few mRNA molecules per cell. MPSS has several drawbacks. First, it is impossible to assess the expression levels because only a limited subset of genes may be assigned signatures.

2.2 Polony Sequencing

Millions of immobilized DNA sequences may be "read" concurrently, using an extremely precise and inexpensive multiplex sequencing technique called polony sequencing. Polony sequencing is often performed on a paired-end tag library, where each DNA template molecule is 135 bp long with two 17–18 bp long paired genomic tags that are isolated from one another and are bordered by common sequences. Current read lengths for this method leave a gap of 4-5 bases between each tag and amplicon, a total of 26 bases per amplicon, and 13 bases per tag. The three main parts of the Polony sequencing process are template amplification, DNA sequencing, and creation of a paired end-tag library.

2.3 Pyrosequencing

Pyrosequencing is a technique for sequencing DNA that works by identifying the nucleotide that a DNA polymerase has absorbed. The "sequencing by synthesis" theoretical paradigm serves as its foundation. When pyrophosphate is released, pyrosequencing employs a chain reaction to detect light. As a result, "pyrosequencing." By combining information from the detection of individual nucleotides added to the developing DNA with the light generated by luciferase, pyrosequencing produces sequence readouts. In contrast to Solexa and SOLiD on the one hand and Sanger sequencing on the other, this method offers an intermediate read length and price per base.

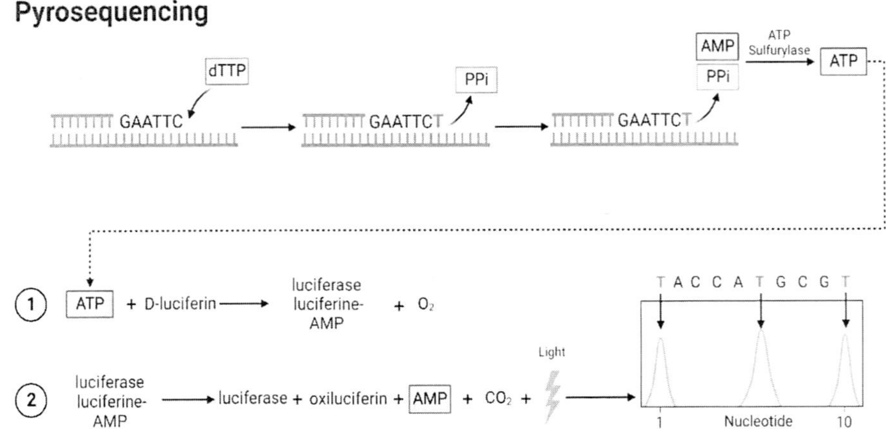

Fig. 2. Schematic representations of Pyrosequencing workflows (Biorender.com)

2.4 DNA Nanoball Sequencing

High-throughput sequencing technology can be used to determine the whole genomic sequence of an organism. Compared to other next-generation sequencing methods, DNA sequencing allows the sequencing of DNA nanoballs on a massive scale at a relatively low cost per run of reagents. Rolling circle replication is used to amplify genomic DNA fragments. However, only short DNA sequences are produced by each DNA nanoball, making it more difficult to match short reads to a reference genome. Numerous genome sequencing studies have used this technology and more studies are planned.

2.5 Single Molecule Sequencing using Helioscope

Helioscope sequencing involves the use of DNA fragments with polyA tail adapters that are inserted and attached to the flow cell surface. The next steps involved cleaning the flow cell cyclically with fluorescently labeled nucleotides and then performing extension-based sequencing. The Helioscope sequencer performs short reads that can consist of up to 55 bases in each run. However, recent advances in this approach have made it possible to obtain more precise reads for homopolymers and RNA sequencing.

2.6 Single Molecule SMRT Sequencing

SMRT sequencing is based on sequencing using a synthetic approach. DNA synthesis takes place in small, well-like containers called zero-mode waveguides (ZMWs). The capture instruments were positioned at the bottom of the wells. Fluorescent nucleotides floating freely in the solution and unmodified polymerase were used for the sequencing. Nucleotide changes were detected using SMTR. This was enabled by the observation of polymerase kinetics. Using this method, up to 1000 nucleotides can be read.

Workflow of next-gen sequencing

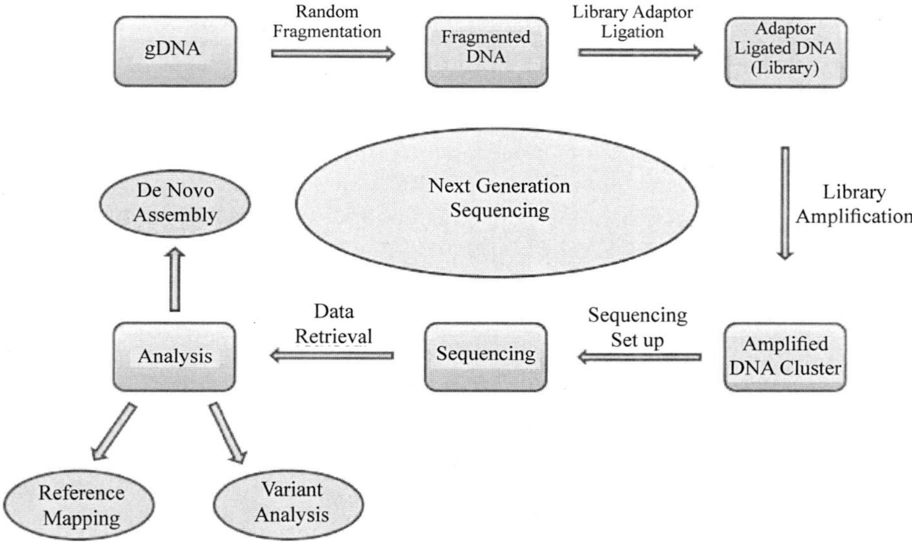

Fig. 3. NGS detailed workflow

3. Applications of NGS in Aquaculture and Fisheries

3.1 Water quality assessment

The NGS-based techniques are applicable to those researches that are going on the water quality and its assessment. NGS was used to determine the appropriateness of and biases associated with its use as an analytical tool. It evaluates biological risks and narrates the advantages and disadvantages of NGS data interpreted using direct quantitative analysis. Studies have been conducted on the distribution of genes linked to antibiotic resistance and toxicity in water samples, as well as the biodegradation processes of harmful contaminants that compromise water quality. Using NGS-based techniques, bioindicators for sewage contamination and microbiological source tracing have been developed.

Although Next-Generation Sequencing (NGS) has shown promise in identifying signs of human fecal or sewage contamination to monitor water quality degradation, a more straightforward strategy for assessing biological risks in water involves directly assessing the variety and prevalence of pathogens.

However, tracking aquatic infections individually is costly, technically difficult, and requires determining which diseases to target beforehand (Varela and Manaia, 2013). The ability to distinguish taxonomically above the level of a family or genus is hampered by this short length. However, as Next-Generation Sequencing (NGS) advancements have extended sequence read lengths, it is now possible to confidently classify sequences produced by NGS at the bacterial species level.

3.2 NGS in the routine clinical microbiological diagnostic laboratory

An appropriate treatment method for infectious diseases can be established through the detection, identification, and characterization of pathogens. Therefore, a wide range of culture-dependent and-independent techniques are available in routine clinical microbiological diagnostic laboratories to identify the source of microbial infections. Numerous organisms require particular growth conditions, which are sometimes unable to adapt to the laboratory setting, making it impossible to identify the causal agent using culture-dependent methods (Lagier *et al.*, 2015). These findings will eventually result in an enhanced comprehension of how diseases develop, paving the way for a new era in personalized medicine and molecular pathology. In microbiological studies, a culture-independent method has been used, but not in routine clinical microbiological diagnostics. Next-generation sequencing (NGS) was performed. The utilization of NGS and its diverse methodological adaptations enable the simultaneous identification of various microorganism types within a microbial sample. This is achieved without relying on traditional culturing methods, and can be accomplished using a single sequencing procedure. (Goodwin *et al.*, 2016).

3.3 NGS as a future authentication tool in food testing

For consumers, regulatory agencies, and those involved in food production and processing, ensuring food authenticity is a major concern. This is due to the fact that instances of incorrect food labelling and other deceptive practices have been shown to reduce consumer trust and, in some cases, even endanger consumer safety (Barnett *et al.*, 2016). Therefore, food product authenticity is a crucial issue for consumers, authorities, farmers, and processors because dishonest business practices can have a negative impact on consumer safety and confidence, as well as the business models of reliable organizations. The use of DNA sequencing is spreading rapidly, which has caused databases that contain sequences to grow quickly in response. In terms of accuracy, creating trustworthy databases might be advantageous for the successful application of NGS. Strong quality control standards and proficiency testing techniques, both of which are currently under development, are required for these applications.

Table 2. Next-generation sequencing technologies and applications (Zhou *et al.*, 2010).

S.No.	Category	Applications	References
1.	Genome	de novo sequencing: the initial generation of large eukaryotic genomes	Li *et al.*, 2010
2.	Transcriptome	measurement of alternative splicing and gene expression; annotation of transcripts; identification of somatic mutations or transcribed SNPs	Jacquier, 2009
3.	Epigenome	1. small RNA profiling 2. Methylation of DNA	Houwing *et al.*, 2007 Costello *et al.*, 2009
4.	Metagenome	genomic profiles of nucleosome positions	Johnson *et al.*, 2006

4. Some other applications of NGS are

1. Mutation discovery
2. Transcriptome analysis- RNA-Seq
3. Enabling metagenosis
4. ChiP-Seq: Defines DNA-Protein Interactions
5. Discovering non-coding RNAs
6. Molecular diagnostics for Oncology and Inherited Disease study
7. Gene regulation analysis
8. Whole genome sequencing
9. Exploring chromatin Packaging

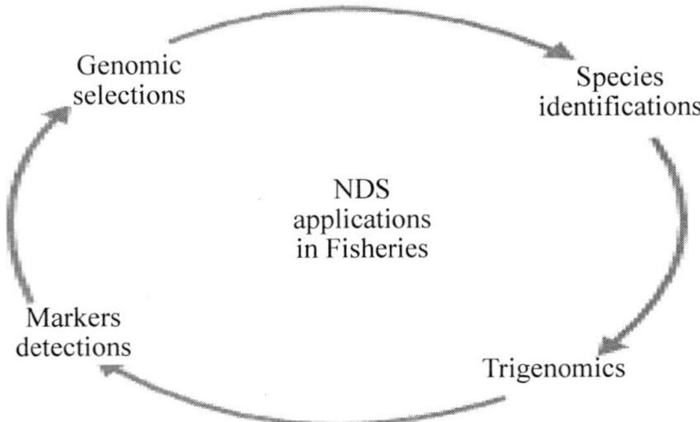

Fig. 4. NGS in Aquaculture

Conclusion and Future Perspectives

The genomic era has changed because of sequencing technologies, which have also assisted us in better understanding and describing the genomes of plants, animals, and humans. The sequencing process is becoming increasingly easy, rapid, and inexpensive because of daily advances in sequencing chemistry, throughput, and nucleotide detection. The fields of molecular biology and genomics have been transformed by next-generation sequencing. It enables faster genome sequencing and annotation as well as genome-wide research on DNA binding, variation, and expression. The least expensive sequencing technology available to date is sequencing by synthesis (Illumina platform), which is also the most popular technology because of its low cost, low error rate, quick turnaround times, and high throughput. This is true, even though the approach has a shorter read length. Newer generations of sequencing technologies, such as nanostructure-based sequencing and single-molecule sequencing, have the potential to develop faster, more affordable, accurate, and dependable methods of producing sequence data. The drawbacks of the present next-generation sequencing methods, such as short-reads and reduced base accuracy, will be solved with the development of new technologies. This is a fascinating field of genetic research.

References

Barnett, J., Begen, F., Howes, S., Regan, A., McConnon, A., Marcu, A., ... & Verbeke, W. (2016). Consumers' confidence, reflections and response strategies following the horsemeat incident. Food Control, 59, 721-730.

Costello, J.F., Krzywinski, M., and Marra, M.A. (2009). A first look at entire human methylomes. Nat Biotechnol 27, 1130–1132.

Goodwin, S., McPherson, J. D., & McCombie, W. R. (2016). Coming of age: ten years of next-generation sequencing technologies. Nature Reviews Genetics, 17(6), 333-351.

Houwing, S., Kamminga, L.M., Berezikov, E., Cronembold, D., Girard, A., van den Elst, H., Filippov, D.V., Blaser, H., Raz, E., Moens, C.B., *et al.* (2007). A role for Piwi and piRNAs in germ cell maintenance and transposon silencing in Zebrafish. Cell 129, 69–82.

https://www.biorender.com/template/pyrosequencing

Illumina Inchttps://www.illumina.com/systems/sequencingplatforms. html.2022.

Jacquier, A. (2009). The complex eukaryotic transcriptome: unexpected pervasive transcription and novel small RNAs. Nat Rev Genet 10, 833–844.

Johnson, S.M., Tan, F.J., McCullough, H.L., Riordan, D.P., and Fire, A.Z. (2006). Flexibility and constraint in the nucleosome core landscape of Caenorhabditis elegans chromatin. Genome Res 16, 1505–1516.

Lagier, J. C., Hugon, P., Khelaifia, S., Fournier, P. E., La Scola, B., & Raoult, D. (2015). The rebirth of culture in microbiology through the example of culturomics to study human gut microbiota. Clinical microbiology reviews, 28(1), 237-264.

Lahens, N. F., Ricciotti, E., Smirnova, O., Toorens, E., Kim, E. J., Baruzzo, G., ... & Grant, G. R. (2017). A comparison of Illumina and Ion Torrent sequencing platforms in the context of differential gene expression. BMC genomics, 18, 1-13.

Li, R.Q., Fan,W., Tian, G., Zhu, H.M., He, L., Cai, J., Huang, Q.F., Cai, Q.L., Li, B., Bai, Y.Q., *et al.* (2010). The sequence and de novo assembly of the giant panda genome. Nature 463, 311–317.

Mostafa, E. M., Sabri, D. M., & Aly, S. M. (2015). Overviews of "next-generation sequencing". Research and Reports in Forensic Medical Science, 1-5.

Niedringhaus, T. P., Milanova, D., Kerby, M. B., Snyder, M. P., & Barron, A. E. (2011). Landscape of next-generation sequencing technologies. Analytical chemistry, 83(12), 4327-4341.

Pervez, M. T., Abbas, S. H., Moustafa, M. F., Aslam, N., & Shah, S. S. M. (2022). A comprehensive review of performance of next-generation sequencing platforms. BioMed Research International, 2022.

Rothberg, J. M., Hinz, W., Rearick, T. M., Schultz, J., Mileski, W., Davey, M., ... & Bustillo, J. (2011). An integrated semiconductor device enabling non-optical genome sequencing. Nature, 475(7356), 348-352.

Rothberg, J. M., Hinz, W., Rearick, T. M., Schultz, J., Mileski, W., Davey, M., ... & Bustillo, J. (2011). An integrated semiconductor device enabling non-optical genome sequencing. Nature, 475(7356), 348-352.

Varela, A. R., and Manaia, C. M. (2013). Human health implications of clinically relevant bacteria in wastewater habitats. Environ. Sci. Pollut. Res. Int. 20, 3550–3569. doi: 10.1007/s11356-013-1594-0.

Zhou, X., Ren, L., Meng, Q., Li, Y., Yu, Y., & Yu, J. (2010). The next-generation sequencing technology and application. Protein & cell, 1(6), 520-536.

4

Future Prospective of eDNA in Aquaculture, its Challenge to Small Farmers

Felegush Erarto[1*], Durgesh Kumar Verma[2], Sanjay Kumar Gupta[3] and Soibam Khogen Singh[4]

[1]*Department of Fisheries and Aquatic Sciences, Bahir Dar University, Ethiopia*

[2]*ICAR-Central Inland Fisheries Research Institute, Barrackpore, West Bengal Regional Centre, Prayagraj-211002, Uttar Pradesh, India*

[3]*ICAR-Indian Institute of Agricultural Biotechnology, Namkum-834010, Ranchi, India*

[4]*Krishi Vigyan Kendra, Ukhrul ICAR RC NEH Region, Manipur, India*

Abstract

Aquaculture is vital for addressing global food security and nutritional needs, contributing to the Sustainable Development Goal (SDG) 2 by providing a sustainable source of protein and essential micronutrients. It also supports SDG 3 by promoting good health and well-being through access to nutritious seafood, and SDG 8 by generating employment opportunities and fostering economic growth, particularly in developing nations. However, aquaculture causes environmental problems and disease outbreaks. These challenges can be addressed by adopting sustainable aquaculture techniques, investing in research, strengthening regulations, monitoring, and raising public awareness. eDNA is a promising and noninvasive tool that helps revolutionize small-scale and large-scale aquaculture practices. It can promote sustainable aquaculture practices by providing valuable insights into aquatic ecosystems, optimizing production, and minimizing environmental impact. Implementing eDNA also requires a comprehensive approach encompassing cost reduction strategies, capacity building, mobile testing solutions, standardization efforts, user-friendly tools, collaborative research, and supportive policies. This also helps overcome the challenges faced by small-scale farmers and enhances the long-term sustainability of the aquaculture sector to meet the world's rising demand for seafood.

Keywords: *Aquaculture, eDNA, Production optimization, Small scale farmers, Sustainability*

1. Introduction

By 2030, sustainable development must be achieved through cross-sector cooperation, creativity and comprehensive initiatives. Aquaculture has contributed to SDGs; however, its contribution depends on the effectiveness of local and international governance (Farmery *et al.*, 2020). Because it includes all aquatic ecosystems, it is one of the global food-producing industries that does so at one of the fastest rates (Mugimba *et al.*, 2021), accounting for approximately half (46%) of the total global fish production (FAO, 2020). Understanding the various environmental, social, and economic aspects of aquaculture offers a more specialized method for handling opportunities and challenges for future development (Troell *et al.*, 2023). Global seafood consumption has risen from 9.0 kg in 1961 to 20.2 kg in 2020, accounting for over 17% of the world's intake of animal proteins (FAO 2022). Indeed, there are considerable regional differences in the production potential and value of the aquaculture sector (Naylor *et al.*, 2021). Over the past 20 years, China, India, Indonesia, Vietnam, Bangladesh, and Egypt have been the major producing countries that have consolidated their share of regional or global production (FAO 2020). Moreover, developing nations account for more than 95% of global aquaculture production, growing at an average annual rate of 6.13 percent (FAO, 2020). As the name aquaculture indicates, it involves farming various aquatic organisms, ranging from the production of unicellular Chlorella algae within indoor bioreactors (Gors *et al.*, 2010) for Atlantic salmon *(Salmo salar)* production in outdoor floating net cages (FAO, 2019). The aquaculture production of fish is the largest, accounting for over half (53.4 million tonnes) of all farmed aquatic species. Aquatic plants (31.8 million tonnes), mollusks (17.4 million tonnes), crustaceans (8.4 million tonnes), amphibians and reptiles (471,784 tons), and miscellaneous invertebrate animals (422,124 tons) are among the other important aquatic species farmed (Tacon 2019). Although aquaculture is a fast-expanding industry that provides billions of people with food and is a means of subsistence worldwide, several obstacles could prevent this sector's expansion and sustainability on a global scale.

Some significant aquaculture challenges include environmental impacts, disease outbreaks, feed availability, access to market and trade barriers, social acceptance and consumer perception, climate change, infrastructure, skilled human resources, and regulations (Mugimba *et al.*, 2021). Therefore, solving these issues is crucial for sustaining the expansion and viability of the aquaculture sector. On the other hand, aquaculture can continue to be vital for the global food and livelihood supply chain by investing in research and development, enhancing aquaculture practices, and fortifying governance and regulation. An innovative technology that promises to resolve many problems facing aquaculture worldwide is environmental DNA (eDNA) (Sahu *et al.*, 2023). It is a rapid, easy, quick, and reasonable scientific technique that has a broad range of applications (Harper *et al.* 2018; Peters *et al.*, 2018; Pawlowski *et al.*, 2020; Bohara *et al.*, 2022b; Sahu *et al.*, 2023). Additionally, eDNA is non-invasive and does not affect any species or habitats being studied, even during sampling (Ramírez-Amaro *et al.*, 2022; Sahu *et al.*, 2023). Fish environmental DNA (eDNA) studies have advanced significantly over the past ten years, and by utilizing this progress, significant strides in fish monitoring have been made (Wang *et al.*, 2021). Environmental DNA (eDNA) monitoring methods have significantly improved fishery management decisions (Bernos *et al.*, 2023). While investigating microbial DNA in sediment samples, Ogram *et al.* originally described environmental DNA (eDNA) (Ogram *et al.*, 1987). This has gained momentum in recent years, driven by an increasing number of studies utilizing DNA from environmental sources (Sahu et. al., 2023). This approach offers significant advantages, including the detection of elusive organisms, conducting large-scale sampling with reduced biases compared to traditional specimen-based methods, and generating valuable data for molecular systematics (Bass *et al.*, 2015). A potential answer to many biological problems is environmental DNA (eDNA), which is the genetic material that is directly retrieved from ambient samples without the requirement for visible biological components Taberlet *et al.,* 2012; Thomsen and Willerslev, 2015; Thomsen *et al.,* 2015, Pawlowski *et al.,* 2020. The eDNA technique has immense potential to revolutionize biodiversity research and aquatic ecosystem conservation. Future advancements are expected to expand eDNA-based approaches beyond single-marker analyses to encompass meta-genomic surveys, thus enabling the prediction of spatial and temporal biodiversity patterns. These developments have broad implications in fields such as biology, geology, and environmental sciences (Peters *et al.*, 2018). (eDNA) water assays are being implemented for many vital pathogens in confined aquaculture systems (Schuster *et al.*, 2023).

2. Future Prospective of eDNA in Aquaculture

While nascent, eDNA technology is rapidly evolving and transforming the aquaculture industry. Its ability to provide efficient, cost-effective, and sensitive monitoring of fish populations, pathogen detection, and invasive species identification has led to many applications. These include assessing fish population health, identifying pathogens and invasive species, tracking fish movement, identifying the source of fish contamination, and developing innovative aquaculture products and services (Sahu *et al.*, 2023). eDNA's potential of eDNA extends beyond aquaculture and encompasses biodiversity research and ecosystem conservation. Future advancements are expected to expand eDNA-based approaches from single-marker analyses to meta-genomic surveys, thereby enabling the prediction of spatial and temporal biodiversity patterns. These developments hold immense promise for aquaculture and other fields such as biology, geology, and environmental science (Thomsen *et al.*, 2015).

2.1 Early detection of disease outbreaks

Diseases are a significant challenge for aquaculture, accounting for approximately 40% of the production losses. It negatively affects survival, growth, reproduction, and market value. The complexity of the interaction between disease outbreaks and water quality measures and the frequency of pathogens in production systems make it challenging to appropriately quantify disease risk. Environmental DNA (eDNA) sampling methodology, combined with molecular techniques, offers a promising solution for rapidly evaluating the presence of pathogens and mitigation measures in fish farms. This approach has demonstrated its potential for assessing parasite loads in water, potentially reducing the risk of disease outbreaks (Bastos Gomes *et al.*, 2017). By analyzing water samples or sediment, eDNA can identify the presence of pathogens even before clinical signs of disease appear in farmed fish populations. This early warning system allows for timely intervention and disease control measures, reducing the spread of infection and minimizing losses to aquaculture production. However, the reliability of eDNA results can be hampered by nucleic acid degradation caused by several factors. Despite these challenges, eDNA has emerged as a valuable tool for detecting various diseases and parasites in aquaculture, as highlighted by several studies (Bohara *et al.*, 2022a). In freshwater aquaculture, parasites are frequently regarded as less important; however, they can still lower the value of the final product. Furthermore, several instances of co-infection with viruses and bacteria have been caused by parasite infestation (Xu *et al.*, 2007).

Gyrodactylus parasites attach themselves to their host's skin, fins, and gills, and feed on the host's blood and tissue, leading to chronic gyrodactylosis (Klesius and Rogers 1995). They are tiny, worm-like individuals that measure only a few millimeters in length and exhibit viviparity, giving birth to live offspring that may already contain a developing embryo. This characteristic, coupled with high birth rates and direct transmission between hosts, enables Gyrodactylus to quickly expand its population (Rawson and Rogers 1973). Gyrodactylus outbreaks, known as epizootics, can result in substantial fish losses and are commonly found in young fish, as they are more susceptible to infection (Buchmann and Lindenstrom 2002). However, they can infect adult fish, particularly if they are stressed or unhealthy. A few steps can be taken to prevent and control Gyrodactylus parasites. One is to keep fish populations healthy and stress-free, and this can be achieved by providing them with a clean and well-maintained environment and a diet high in nutrients. It is also important to avoid overcrowding in fish, as this can increase the risk of infection. If Gyrodactylus parasites are detected in a fish population, few treatment options are available. These include using medications, such as formalin, potassium permanganate, or heat treatment. Heat treatment involves raising the water temperature to a lethal level for parasites, but not for fish. Gyrodactylus parasites can pose a severe problem for fish farmers and hobbyists. However, the spread of these parasites can be controlled by taking preventive measures and promptly treating the infected fish.

Saprolegniasis, branchiomycosis, and aspergillosis are the most common fungal diseases affecting fish species. The high density and constant stress levels experienced by farmed animals and exposure to various pollutants increase the risk of infection and disease spread in fish farms compared with wild environments. Rainwater flow and infiltration facilitated the spread of fungi to ponds. Implementing efficient eDNA-based methods to detect fungal pathogens will enable fish farmers to anticipate fungal disease outbreaks and implement timely management and control strategies. As it holds significant potential for disease risk monitoring, it enhances our ability to determine the presence, diversity, and abundance of pathogenic organisms (Bohara *et al.*, 2022a). However, traditional pathogen detection techniques in aquaculture are time-consuming and resource-intensive.

Freshwater aquaculture is also affected by bacterial infections, which cause significant financial losses to farmers. An effective method for forecasting the number of bacteria in fish farms is to use eDNA methods. For example, regional pathogen profiles were successfully defined over a year using an aquaculture bacterial pathogen database in conjunction with eDNA metabarcoding monitoring data from coastal and mariculture waters in Hong Kong. This has improved the process of identifying possible novel targets for pathogens. The potential of aquaculture bacterial pathogen databases to supplement eDNA-based methods for coastal marine pathogen surveillance has worldwide ramifications for the development of aquaculture and water resource management (Lo *et al.*, 2023).

2.2 Monitoring Environmental Impacts

Aquaculture practices can have both beneficial and detrimental effects on the environment. eDNA monitoring can track these effects by examining the diversity and abundance of aquatic species. For example, eDNA can detect the presence of invasive species that may threaten farmed fish by competing for resources or preying on them. This information can then be used to implement the targeted control measures. Additionally, eDNA can be used to assess the impact of aquaculture activities on sensitive species, enabling the implementation of mitigation strategies to minimize environmental harm. Furthermore, eDNA can aid in optimizing feed management practices and stocking densities, ensuring that farmed fish do not compete with wild populations for food or space.

2.3 Improving genetic management and stock enhancement

In aquaculture, eDNA can be used to monitor the genetic diversity of farmed fish populations, allowing for selective breeding and stock enhancement programs. By identifying genetically superior individuals or populations, eDNA can guide breeding strategies to improve disease resistance, growth rates, and other desirable traits. Genetic management can enhance the sustainability of aquaculture by reducing reliance on artificial inputs and ensuring the long-term health of farmed fish stocks. Integrating eDNA into aquaculture practices can contribute significantly to the sustainability of an industry. By enabling early detection of diseases, monitoring environmental impacts, optimizing feed management, and improving genetic management, eDNA can help reduce the environmental footprint of aquaculture, enhance the health and resilience of farmed fish populations, and promote long-term viability of the industry.

3. Challenges of eDNA in small-scale aquaculture farmers

Despite the numerous benefits of eDNA technology for aquaculture, there are some challenges that small-scale aquaculture farmers may face in implementing and utilizing this tool effectively. However, the application of eDNA in small-scale aquaculture farms faces several challenges (Kelly and Port 2014; Shaw *et al.,* 2018, Adams and Pace, 2020; Bustos-Martínez and Savidge 2020; Doi *et al.* 2020).

3.1 Cost of eDNA analysis: eDNA analysis can be relatively expensive, especially for small-scale aquaculture operations with limited financial resources. The cost of equipment, reagents, and specialized personnel can pose a barrier to entry for small-scale farmers.

3.2 Technical Expertise Requirements: eDNA analysis involves complex laboratory procedures and data interpretation and requires specialized training and expertise. Small-scale farmers may require more technical expertise or access to training opportunities to conduct eDNA analysis independently.

3.3 Interpretation of eDNA Results: Interpreting eDNA results can be challenging because of environmental variability, DNA degradation, and potential false positives or negatives. Small-scale farmers may require support from experts or specialized laboratories to accurately interpret eDNA data and draw meaningful conclusions.

3.4 Integration into Existing Practices: Integrating eDNA monitoring into existing aquaculture practices requires careful planning and adaptation. Small-scale farmers may need to adjust their routines and protocols to accommodate eDNA sampling, data analysis, and decision making based on eDNA results.

3.5 Accessibility of eDNA services: eDNA analysis services may be limited in certain regions, mainly for small-scale aquaculture operations in remote areas. This can create additional challenges in obtaining timely and cost-effective eDNA testing (Adams and Pace 2020).

3.6 Standardization of eDNA Methods: The standardization of eDNA methods and protocols is still ongoing, leading to variability in results and making it difficult for small-scale farmers to compare data across different sources. In addition, the small volume and dynamic nature of aquaculture ponds can lead to variability in eDNA concentrations, making it challenging to obtain representative samples (Bustos-Martínez and Savidge, 2020).

3.7 Education and Awareness: Raising awareness and providing education on eDNA technology among small-scale aquaculture farmers is crucial for its widespread adoption. This can help farmers understand the benefits of eDNA, its limitations, and how to effectively utilize it in their operations.

3.8 Financial Support and Incentives: Government agencies and industry organizations can support small-scale aquaculture farmers by providing financial assistance, training opportunities, and incentives to adopt eDNA technology. This can help to overcome the financial and technical barriers faced by small-scale farmers.

3.9 The development of user-friendly tools and kits specifically designed for small-scale aquaculture operations can make eDNA analysis more accessible and technically less demanding. This can empower small-scale farmers to independently conduct basic eDNA testing.

3.10 Collaboration and Knowledge Sharing: Encouraging collaboration and knowledge sharing among small-scale aquaculture farmers, researchers, and extension services can facilitate the dissemination of eDNA best practices and promote effective utilization of this technology in small-scale aquaculture settings.

4. Overcoming the challenges Small-scale Aquaculture Farmers

Overcoming the challenges of eDNA in aquaculture requires a multifaceted approach that addresses the technological, economic, and educational aspects.

Cost reduction is a potential strategy to address the challenges faced by small-scale aquaculture farmers, and can be achieved by developing cost-effective eDNA extraction and analysis kits specifically designed for small-scale aquaculture applications. Utilizing portable and low-cost eDNA analysis devices that can be operated directly at aquaculture sites and alternative eDNA sampling methods, such as passive sampling devices, to reduce labor costs is essential. In addition, promoting the benefits and applications of eDNA technology among small-scale aquaculture communities through outreach programs and demonstrations, encouraging collaboration and knowledge sharing among small-scale farmers and researchers, availability of eDNA protocols for specific aquaculture, and establishment of eDNA reference centers or laboratories that can provide guidance and support to small-scale farmers are very important to fill the gap in expertise and standardization. Interpretation and Validation can also be overcome by developing user-friendly software and online platforms that can assist small-scale farmers in interpreting eDNA results, and by providing precise and concise interpretation guidelines for eDNA results to aid in decision-making processes. By addressing these challenges, eDNA could become a more accessible and valuable tool for small-scale aquaculture farmers, enabling them to improve their practices, enhance productivity, and ensure sustainability.

5. Future Prospective of eDNA in small- and large-scale aquaculture farmers

Environmental DNA (eDNA) technology is promising for revolutionizing aquaculture practices across all scales, offering a noninvasive, cost-effective, and sensitive tool for monitoring aquatic ecosystems, optimizing production, and promoting sustainable aquaculture. For small-scale aquaculture farmers, eDNA technology can be democratized through cost-effective analysis methods, user-friendly tools, and mobile testing solutions, empowering them to make informed decisions and enhance their practices. Early disease detection and prevention will be facilitated through eDNA monitoring, enabling timely interventions and reducing production losses, thereby improving the overall health and productivity of small-scale aquaculture operations. Additionally, eDNA data will guide feed management strategies and stocking decisions, ensuring that farmed fish do not compete with wild populations for resources or minimize environmental impacts. Furthermore, eDNA can promote sustainable practices by providing insights into ecosystem health, preventing the introduction of invasive species, and minimizing the use of chemicals and antibiotics.

Large-scale aquaculture operations will benefit from eDNA technology by optimizing production and efficiency using valuable data on production parameters, such as feed amounts, stocking densities, and harvesting strategies. This leads to increased efficiency and resource conservation. Maintaining ecosystem integrity will be enabled through eDNA monitoring, allowing large-scale aquaculture operations to assess their impact on surrounding ecosystems, implementing mitigation measures to minimize environmental harm, and promoting coexistence with wild populations. eDNA will also provide an early warning of disease outbreaks, enabling timely intervention and preventing widespread losses. It is also possible to guarantee consumer trust in the sustainability and safety of aquaculture products via eDNA monitoring, which will promote environmental stewardship and ethical operations.

Future advancements in eDNA technology will further enhance its capabilities, including the development of more sensitive and specific eDNA detection methods to detect even lower concentrations of DNA and improve its ability to detect rare or cryptic species and pathogens. Data analysis and interpretation advancements will provide more accurate and meaningful insights from eDNA data, thus providing more actionable information for aquaculture management decisions. Integrating eDNA with other monitoring tools such as sensors and imaging technologies will provide a comprehensive picture of the aquatic environment. Real-time eDNA monitoring can immediately detect changes in aquatic ecosystems, enabling rapid responses to potential threats or disturbances. In conclusion, eDNA technology holds immense potential to transform aquaculture practices and enhance sustainability, productivity, and environmental stewardship for both small-scale and large-scale aquaculture operations. As technology continues to advance and its applications expand, eDNA is vital for enhancing aquaculture and contributing to achieving sustainable development goals.

Conclusion

Generally, environmental DNA (eDNA) monitoring has emerged as a valuable tool in aquaculture, offering a noninvasive and cost-effective approach to assessing species presence and abundance, thereby contributing to the management and maintenance of healthy aquatic environments. Specifically, it assists in controlling water quality by tracking the presence and density of aquatic organisms, allowing it to guide stock management decisions and prevent overpopulation and associated water quality degradation. It also aids in detecting pathogens or diseases carried by aquatic organisms, enabling early intervention to prevent disease outbreaks that could compromise the water quality. Monitoring eDNA levels can also inform feed management strategies, optimize feed amounts, and minimize waste production; it helps maintain water quality and prevents excessive nutrient enrichment. It also provides an understanding of the presence and behavior of different species, facilitating the maintenance of a balanced ecosystem. It prevents the domination of a particular species that may adversely affect water quality and overall ecosystem health. Thus, eDNA monitoring is essential for promoting sustainable practices and protecting the well-being and productivity of aquaculture farms.

References

Adams, N. M., and Pace, M. R. (2020). The Utility of Environmental DNA for Monitoring the Health of Aquaculture Systems. Reviews in Aquaculture, 12(4), 1217-1230.

Bass, D., Stentiford, G. D., Littlewood, D. T. J.,and Hartikainen, H. (2015). Diverse Applications of Environmental DNA Methods in Parasitology. Trends Parasitol, 31(10), 499-513. doi:10.1016/j.pt.2015.06.013

Bastos Gomes, G., Hutson, K. S., Domingos, J. A., Chung, C., Hayward, S., Miller, T. L.,and Jerry, D. R. (2017). Use of environmental DNA (eDNA) and water quality data to predict protozoan parasites outbreaks in fish farms. Aquaculture, 479, 467-473. doi:10.1016/j.aquaculture.2017.06.021

Behringer, D.C., Wood, C.L., Krkošek, M., and Bushek, D., Disease in fisheries and aquaculture In: Marine Disease Ecology. Edited by: Donald C. Behringer, Kevin D. Lafferty, and Brian. R. Silliman, Oxford University Press (2020). © Oxford University Press. DOI: 10.1093/oso/9780198821632.003.0010

Bernos, T. A., Yates, M. C., Docker, M. F., Fitzgerald, A., Hanner, R., Heath, D., Mandrak, N. E. (2023). Environmental DNA (eDNA) applications in freshwater fisheries management and conservation in Canada: overview of current challenges and opportunities. Canadian Journal of Fisheries and Aquatic Sciences, 80(7), 1170-1186. doi:10.1139/cjfas-2022-0162

Bohara, K., Yadav, A. K.,andJoshi, P. (2022a). Detection of Fish Pathogens in Freshwater Aquaculture Using eDNA Methods. Diversity, 14(12). doi:10.3390/d14121015

Bohara, K., Yadav, A. K.,an dJoshi, P. (2022b). Detection of fish pathogens in freshwater aquaculture using eDNA methods: a review. Preprints (www.preprints.org) doi:10.20944/preprints202210.0291.v1

Busch S., Dalsgaard I. and Buchmann K. (2003) Concomitant exposure of rainbow trout fry to Gyrodactylus derjavini and Flavobacterium psychrophilum: effects on infection and mortality of host. Veterinary Parasitology 117, 117–122.

Bustos-Martínez, H.,and Savidge, J. A. (2020). Environmental DNA for Biodiversity Monitoring in Freshwater Ecosystems: A Review. Journal for Nature Conservation, 58, 106-114.

Doi, H., Takahara, T.,and Yamanaka, T. (2020). Application of Environmental DNA for Monitoring Biodiversity and Disease in Aquatic Ecosystems. In Environmental DNA Metabarcoding (pp. 229-265). Springer, Cham.

Kelly, R. P.,and Port, J. A. (2014). Environmental DNA: A New Tool for Biomonitoring. Marine Pollution Bulletin, 88(1-4), 28-35.

Klesius P. and Rogers W. (1995) Parasitisms of catfish and other farm raised food fish. Journal of the American Veterinary Medical Association 207, 1473–1478.

Lo, L. S. H., Liu, X., Liu, H., Shao, M., Qian, P. Y., and Cheng, J. (2023). Aquaculture bacterial pathogen database: Pathogen monitoring and screening in coastal waters using environmental DNA. Water Res X, 20, 100194. doi:10.1016/j.wroa.2023.100194

Miya, M. (2022). Environmental DNA metabarcoding: a novel method for biodiversity monitoring of marine fish communities. Annual Review of Marine Science, 14, 161-185.

Mugimba, K. K., Byarugaba, D. K., Mutoloki, S., Evensen, O., and Munang'andu, H. M. (2021). Challenges and Solutions to Viral Diseases of Finfish in Marine Aquaculture. Pathogens, 10(6). doi:10.3390/pathogens10060673

Ogram A., Sayler G. S., Barkay T. (1987). The Extraction and Purification of Microbial DNA from Sediments. Journal of Microbiological Methods 7(2-3): 57-66.

Pawlowski, J., Apotheloz-Perret-Gentil, L.,andAltermatt, F. (2020). Environmental DNA: What's behind the term? Clarifying the terminology and recommendations for its future use in biomonitoring. Mol Ecol, 29(22), 4258-4264. doi:10.1111/mec.15643

Peters, L., Spatharis, S., Dario, M. A., Dwyer, T., Roca, I. J. T., Kintner, A.,. Praebel, K. (2018). Environmental DNA: A New Low-Cost Monitoring Tool for Pathogens in Salmonid Aquaculture. Front Microbiol, 9, 3009. doi:10.3389/fmicb.2018.03009

Pietramellara G, Ascher J, Borgogni F et al. (2009) Extracellular DNA in soil and sediment: fate and ecological relevance. Biology and Fertility of Soils, 45, 219–235.

Ramírez-Amaro, S., Bassitta, M., Picornell, A., Ramon, C., and Terrasa, B. (2022). Environmental DNA: State-of-the-art of its application for fisheries assessment in marine environments. Frontiers in Marine Science, 9. doi:10.3389/fmars.2022.1004674

Rawson M.V. and Rogers W.A. (1973) Seasonal abundance of Gyrodactylus macrochiri Hoffman and Putz, 1964 on bluegill and largemouth bass. Journal of Wildlife Diseases 9, 174-177

Sahu, A., Kumar, N., Pal Singh, C.,andSingh, M. (2023). Environmental DNA (eDNA): Powerful technique for biodiversity conservation. Journal for Nature Conservation, 71. doi:10.1016/j.jnc.2022.126325

Schuster, C. J., Murray, K. N., Sanders, J. L., and Kent, M. L. (2023). Application of an eDNA assay for the detection of Pseudoloma neurophilia (Microsporidia) in zebrafish (Danio rerio) facilities. Aquaculture, 564. doi:10.1016/j.aquaculture.2022.739044

Shaw, S. G., Petit, E. J.,and Sato, K. (2018). Environmental DNA for Monitoring the Success of Wetland Restoration. Restoration Ecology, 26(3), 567-576.

Taberlet P., Prud'homme S. M., Campione E., Roy J., Miquel C., et al. (2012). Soil sampling and isolation of extracellular DNA from large amount of starting material suitable for metabarcoding studies. Molecular Ecology 21(8): 1816-1820.

Tacon, A. G. J. (2019). Trends in Global Aquaculture and Aquafeed Production: 2000–2017. Reviews in Fisheries Science and Aquaculture, 28(1), 43-56. doi:10.1080/23308249.2019.1649634

Thomsen, P. F., and Willerslev, E. (2015). Environmental DNA – An emerging tool in conservation for monitoring past and present biodiversity. Biological Conservation, 183, 4-18. doi:10.1016/j.biocon.2014.11.019

Thomsen P. F. and Willerslev E. (2015). Environmental DNA - An emerging tool in conservation for monitoring past and present biodiversity. Biological Conservation 183: 4-18.

Troell, M., Costa-Pierce, B., Stead, S., Cottrell, R. S., Brugere, C., Farmery, A. K.,. Barg, U. (2023). Perspectives on aquaculture's contribution to the Sustainable Development Goals for improved human and planetary health. Journal of the World Aquaculture Society, 54(2), 251-342. doi:10.1111/jwas.12946

Wang, S., Yan, Z., Hanfling, B., Zheng, X., Wang, P., Fan, J.,andLi, J. (2021). Methodology of fish eDNA and its applications in ecology and environment. Sci Total Environ, 755(Pt 2), 142622. doi:10.1016/j.scitotenv.2020.142622

Xu, D.-H.; Shoemaker, C.A.; Klesius, P.H. Evaluation of the Link between Gyrodactylosis and Streptococcosis of Nile Tilapia, Oreochromis Niloticus (L.). J. Fish Dis. 2007, 30, 233–238

5

An Insights into the Biofloc Fish Farming Status of India

Magdeline Christo*[*] *and Ashwin Sivasankar

Department of Aquaculture, Kerala University of Fisheries and Ocean Studies Kochi-682506, Kerala, India

Abstract

Aquaculture is one of the rapidly growing sectors of food production, and an increase in the global population has paved the way for a sharp rise in the demand for food fish. Consequently, aquaculture stands out as a food production sector that presents opportunities to address issues related to hunger, malnutrition, and poverty. However, challenges arise in establishing a sustainable aquaculture industry because of the restricted access to natural resources and their adverse effects on the environment. To overcome these obstacles and boost production, it is essential to embrace innovative technologies such as intensive aquaculture with restricted water exchange. An encouraging solution to these challenges is the adoption of biofloc-based culture. Biofloc technology, or BFT, is recognized as an emerging cultivation system with significant potential to culture fish. Biofloc technology is based on the concept of nutrient cycling via intricate biological pathways to generate natural food for fish. Utilizing biofloc in several fish species presents several benefits, such as increased production due to enhanced performance on growth and survival, as well as improved physiological activity in fish aquaculture. This book chapter provides insights into the establishment of the Biofloc system, major cultivable fish species, an overview of production, current status, advantages, applications, limitations, and challenges.

Keywords: *Biofloc system, Sustainable aquaculture, Current Status, Production, Challenge*

1. Introduction

The aquaculture industry is a rapidly expanding domain of food production worldwide. With the exponential growth of the human population, there has been a corresponding rise in the demand for natural resources. Therefore, all technologies aim to maximize food production. The widespread expansion of aquaculture, especially in coastal areas, increases the load of biomass in the culture water and presents enduring ecological hazards. (Piedrahita 2003; Sharifinia *et al.*, 2018, 2019).

The primary challenge confronted by aquaculturists as a result of rapid intensification of the system is the quality of water. Consequently, there has been a surge in eco-friendly management and cultivation practices. Additionally, the development of aquaculture is limited by high land costs and its heavy reliance on fish oil and fish meal (Browdy *et al.*, 2001; De Schryver 2008). Nevertheless, the sector continues to exhibit limited productivity, which is primarily attributed to inadequate quality inputs and substandard culture technologies. With substantial operating and maintenance expenses, the acceptance of recirculating systems among farmers, specifically in developing nations, is limited. Hence, there is a growing requirement for an affordable, environmentally friendly technology that is embraced by farmers and can be widely implemented (Ahmad *et al.*, 2017; Karimanzira *et al.*, 2017). There is growing interest in enclosed aquaculture systems, primarily driven by the biosecurity, environmental, and marketing benefits they offer compared to traditional, semi-intensive, and extensive systems (Ray, 2012).

The eco-friendly aquaculture method known as "Biofloc Technology (BFT)" is commonly acknowledged as a highly effective alternative system, as it allows continuous recycling of nutrients and nutrient reuse. This sustainable approach relies on cultivating microorganisms within the culture, with minimal to no water exchange.

This technology is an environmentally conscious approach that relies on the abundant production of microorganisms on-site. Biofloc floccule refers to a small, loosely aggregated assembly of materials (such as unconsumed feed, waste, and detritus) and affixed organisms (viz., bacteria, fungi, nematodes, protozoa, and phytoplankton) that remain suspended in water within a pond or tank because of vigorous water agitation, which plays an important role in (I) preserving optimal water conditions (Emerenciano *et al.*, 2017), (ii) enhancing the capability of aquaculture through decreased feed conversion ratios (FCR) (Avnimelech, 2012), lower feed expenses, and (iii) ensuring bio-security (Defoirdt *et al.*, 2004); and (iv) capturing and storing greenhouse gases (Manan *et al.*, 2016). The microorganisms utilized in this system have two main roles: first, they help in maintaining water quality by absorbing nitrogen compounds and producing microbial proteins directly on-site; second, they play a nutritional role, enhancing the feasibility of the culture by lowering the FCR and thus reducing feed costs. In the BFT farming method, minimal water release and its reuse serve to avoid environmental degradation, transforming the system into a genuinely "environmentally friendly" one with a "green" approach. The functioning of this technology is basically the concept of recycling nutrients, which is accomplished by a high carbon/nitrogen (C/N) ratio, which exceeds 15, to encourage significant bacterial proliferation (Avnimelech 1999).

An additional source of carbon is introduced into the culture system to enhance the quality of the water, ensuring increased aeration, thereby leading to the growth of microbacterial floc (Crab *et al.*, 2012; Ahmad *et al.*, 2017). Restricted exchange of water aids in sustaining a stable temperature, preventing variations, and facilitating tropical species, even in colder regions (crab *et al.*, 2009). Currently, BFT is known by various alternative names, including zero water Exchange Autotrophic-Heterotrophic Systems (Burford *et al.*, 2004; Wasielesky *et al.*, 2006), suspended-growth systems (Hargreaves 2006), and microbial-floc systems (Avnimelech 2007; Ballester *et al.*, 2010) etc.

2. Establishment of BFT systems

Low-budget biofloc tanks can be created using a wire-mesh framework with a lining of plastic or solid concrete walls. For tanks, ideally 13 m in diameter with an altitude of 1.5 meters, a leveled area of 200–250 sq. ft is necessary. During construction, an iron mesh of 3-4 mm mesh-size was placed on a two-liner framework of bricks laid on the ground. A 20-22° slope is formed from the edge towards the midpoint, where a central outlet pipe is installed for effective sludge drainage. Aeration pipes were then incorporated into the system with multiple outlets for air stones. The tank is filled with water; if necessary, a suitable material is used to provide shade.

Regarding pond setup, conventional earthen ponds lacking liners can be converted into biofloc systems to enhance output. This is accomplished under specific conditions: (1) the orientation of the pond should encourage wind flow to facilitate efficient water mixing; (2) at the inlet location, construct a 2 cubic metre structure using bamboo, which is consistently filled with biodegradable waste to promote the proliferation of worms, bacteria, and plankton-direct fish food sources; and 3) the farmer must maintain an improved C/N ratio by incorporating carbon sources to encourage bacterial proliferation. Alternatively, an increased carbon–nitrogen ratio could be obtained by using feed with a low crude protein content, such as 18%.

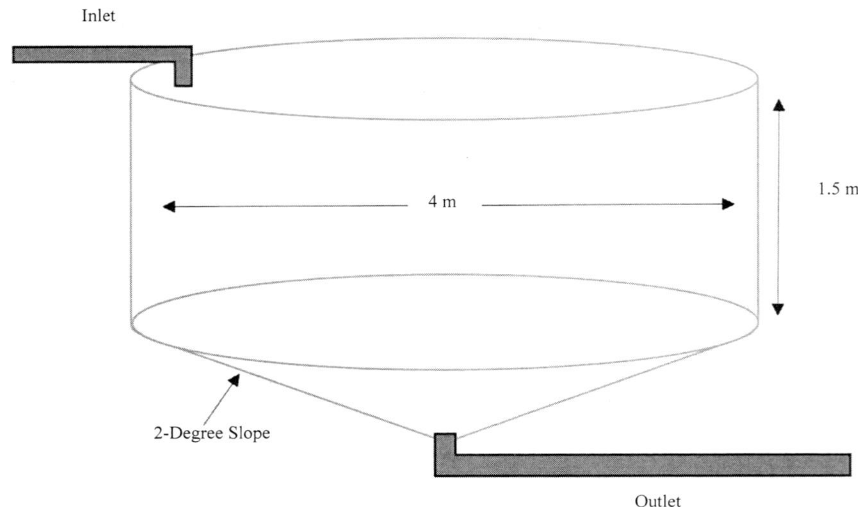

Fig. 1. Layout of Biofloc Tank

3. The Major cultivable fish species in BFT

The fundamental consideration in establishing a biofloc fish farming system involves selecting appropriate species for cultivation. The efficiency is optimized when utilized with species capable of obtaining nutritional advantages through successful utilization of floc. This is particularly well suited for species that can endure increased solid concentrations in water and typically demonstrate resilience to less than optimal water quality conditions. Examples of species suitable for this technology include the following.

- *Clarias batrachus, Anabas testudineus, Ompok pabda, Heteropneustes fossilis, Pangasianodan hypophthalmus.*
- *Cyprinus carpio, Chanos chanos, Labeo rohita, and Oreochromis niloticus.*

  Among these, tilapia are of significant importance, and have gained global recognition for farming purposes. Its successful cultivation in bioflocs can be attributed to its unique characteristics, including its ability to consume particles, tolerance for moderate oxygen levels, ability to thrive in high-density captivity, and adaptability to diverse environmental conditions.
- Shellfishes, such as Penaeus monodon and *Litopenaeus vannamei.*

4. An Overview of Production

In recent years, various studies have been conducted to explore the incorporation of the latest technologies into culture practices such as multi-species aquaculture, recirculation systems, hybrid aquaculture, aquaponics, and the BFT, which aims to enhance overall production.

However, BFT technology faces challenges in garnering support and implementation from farmers because of the absence of flawless technology compared to traditional methods. Conversely, factors such as water scarcity, expensive water resources for aquaculture development, adverse effects of aquaculture effluents on the environment, water pollution, and the risk of infectious diseases have led to a decrease in water exchange in farms, emphasizing the need for a facility with enhanced farm biosecurity (Avnimelech 2009).

It is crucial to impart practical knowledge regarding the successful application and economic advantages of this technology to farmers. The diligent monitoring and examination of ponds is crucial for incorporating BFT into culture. This encompasses monitoring water quality, which includes assessing and stabilizing total suspended solid concentrations and settling solids, and determining the number, type, and placement of aerators in the culture ponds (De Schryver *et al.*, 2008).

Future studies should focus on elucidating the significance and role of biofloc technology in motivating farmers to adopt this system. Customers should also be encouraged to support organic aquaculture products from biofloc ponds. There is a concern that recycling feces into aquatic feed might discourage consumers from purchasing such products. However, with the expanding population, it becomes imperative to develop aquaculture strategies to preserve wild fish population and regulate the prices of marketable fish, as population growth, seafood shortages, and stress on fish population contribute to rising prices of fish (Jiang 2010; Peron *et al.*, 2010).

In contrast to fishery strategic plans, the emphasis is on preserving fish stocks, lowering the prices of fish, and increasing marketable fish stocks. In this situation, biofloc technology holds a promising solution to ease the pressure on aquatic resources, contributing to social well-being by lowering fish production costs. This proves to be advantageous for both farmers and consumers. Consumers seek assurance that the fish they consume are safe for their health, and align with social and ethical considerations. Biofloc technology has been successful in meeting these criteria. Numerous researchers have explored the integration of BFT with other aquaculture techniques to regulate water quality and its associated impacts.

Integrating periphyton with the Biofloc system (Asaduzzaman *et al.*, 2008), merging autotrophic and heterotrophic consortium in the culture unit for environmental regulation (Avnimelech 2009), and exploring the usage of nitrification, denitrification and anaerobic oxidation of NH^3 for removal of nitrogen have been subjects of investigation (Kumar and Lin, 2010). The findings from the aforementioned studies indicate that biofloc technologies with lower energy consumption have the capacity to effectively manage detrimental substances in aquaculture units.

In addition, researchers have explored the application of biofloc technology for multi-species aquaculture, such as combining shrimp with microalgae, tilapia with vegetables, and mussels with seaweed, yielding good outcomes (Kuhn *et al.*, 2009). The integration of biofloc technology into multi-species ponds has resulted in improvements in the quality of water, availability of natural feed, enhanced nutritional performance, and overall production and growth (Rahman *et al.*, 2008).

The adoption of biofloc technology highlights aquaculture as an environmentally sound solution, emphasizing the incorporation of social, economic, and environmental factors into its advancement. Ongoing efforts by researchers aim to further refine this method and encourage its implementation, particularly in the context of new-generation aquaculture. However, the careful adjustment and implementation of this technology necessitates further investigation and data collection to create a strong platform for sustainable aquaculture, involving collaboration among researchers, farmers, and consumers.

5. Current Status of BFT

The aim of Biofloc Technology is to efficiently control and sustain optimal water quality by utilizing the microbial absorption of metabolic byproducts. This approach aims to minimize waste discharge and provide a supplementary source of food for aquaculture species, thereby lowering expenses associated with costly and unnecessary filtration systems and lowering feed costs. The adoption of the BFT approach is increasingly prevalent in farming operations, notably in the U.S. and Mexico, for shrimp and tilapia cultivation. In nations such as Singapore, Vietnam, Thailand, and Indonesia, Biofloc technology is more widely utilized in shrimp culture because of its higher profitability.

A noteworthy benefit of Biofloc Technology is the generation of distinct amino acids that stimulate taste and smell in aquaculture species. This stimulation facilitated an accelerated growth rate and reduced food wastage. Studies have shown that amino acids such as glutamate, glycine, alanine, and arginine present in bioflocs actively capture the interest of shrimp, enticing them to feed (Ju *et al.*, 2008; Nunes *et al.*, 2006). Additionally, highly intensive biofloc culture

culture of white-leg shrimp *(Litopenaeus vannamei)* has been demonstrated to significantly enhance the uptake of phosphorus (P) and nitrogen (N) by 66% and70%, respectively, in comparison with conventional recirculatory systems employing water exchange methods (Da Silva *et al.*, 2013).

As previously stated, waste metabolites, such as phosphates and nitrates, are generated and recirculated within the biofloc system, where they are absorbed again and integrated into the biomass of microorganisms. While biofloc holds promise in controlling nitrate and phosphate levels, its distribution, particularly on a large marketable scale, has both advantages and challenges. A thorough analysis of the strengths, weaknesses, opportunities, and threats is necessary to determine the viability of a commercial farm. The decision to implement BFT is contingent on individual farmers' specific needs, but appealing factors may include reduced water wastage, lower Feed Conversion Ratios (FCR), and increased productivity, addressing potential water and sustainable feed shortages.

6. Advantages and Applications of BFT

The benefits of utilizing BFT in aquaculture have been adequately documented, including economic benefits such as a decreased need for feed and water. Additionally, it mitigates the risk of introducing pathogens and diseases, enhances biosecurity measures, and results in elevated growth and survival rates, consequently leading to increased crop yields (Otoshi *et al.*, 2009; Crab *et al.* 2009; Krummenauer *et al.*, 2011; Perez-Fuentes *et al.*, 201). Furthermore, it diminishes the FCR by using natural feed present in the environment and has a compact footprint, thereby minimizing the environmental impacts. This approach is well known for its robustness, ease of operation, and economic feasibility.

7. Challenges and Limitations of BFT

Although there are many benefits associated with the BFT, certain factors hinder its efficient implementation. While promoting a higher bacterial load, the excessive growth of heterotrophic bacteria leads to increased turbidity in the system. This turbidity, in turn, may result in clogging of the gills of shrimp and fish, particularly in fish species that are not well-suited to thriving in murky waters. The BFT method demands increased energy usage for aeration and mixing, posing a challenge in regions with an unstable power supply marked by frequent blackouts, a common issue in developing nations.

The BFT process also requires a starting period of approximately two weeks for microbial development, potentially extending production cycles. Alkalinity supplementation is important for maintaining a favorable environment for the proliferation of bioflocs. An imbalance in microbial populations, such as extremely low bacterial levels, can elevate the risk of pollution owing to the buildup of nitrate compounds. Additionally, systems exposed to sunlight may experience inconsistent and seasonal performances, requiring production halts during heavy rainy seasons. Further obstacles include the potential outperformance of filamentous flocs, leading to inadequate nitrogen elimination through floc bulking, finally leading to system instability.

Conclusion

Water pollution is a significant concern due to the growing demand for seafood, along with other limitations on land for expanding aquaculture operations in the aquaculture sector. Intensive aquaculture has emerged as a primary solution for addressing the rising demand for animal proteins. Achieving sustainable development and environmentally friendly aquaculture is possible through the adoption of BFT. Incorporating bioflocs into our culture system is crucial, as it addresses the supplementation needs of aquatic organisms and functions as a viable substitute for supplementary feed in their diet.

Bioflocs contain microbial proteins as well as bacterial consortia with peptidoglycans and lipopolysaccharides. When integrated with commercial diets, bioflocs establish a holistic food chain that sustains aquatic development and boosts the overall efficiency. Research indicates that the presence of bioflocs in zero water exchange systems paves the way for better water quality and increased rate of growth in species such as Pacific white shrimp, as they actively utilize these particles. By leveraging the nutritional value of bioflocs, their application in a moist state within a similar culture area, and the use of dried biofloc incorporated into aquatic feed can efficiently decrease the cost of feed and subsequently lower production expenses. Thus, it is evident that standardized techniques, methods, pond construction equipment, stocking management, and harvest in biofloc technology-based aquaculture systems.

References

Ahmad, I., Babitha Rani, A.M., Verma, A.K., & Maqsood, M. (2017). Biofloc technology: an emerging avenue in aquatic animal healthcare and nutrition. Aquaculture International 25,1215–1226.

Asaduzzaman, M., Wahab, M., Verdegem, M., Huque, S., Salam, M., & Azim, M. (2008). C/N ratio control and substrate addition for periphyton development jointly enhance freshwater prawn Macrobrachium rosenbergii production in ponds. Aquaculture, 280(1-4), 117–123.

Avnimelech ,Y. (2007). Feeding with microbial flocs by tilapia in minimal discharge bioflocs technology ponds. Aquaculture, 264(1-4),140–147.doi:10.1016/j.aquaculture. 2006. 11.025

Avnimelech, Y. (2009). Biofloc Technology: A Practical Guide Book. World Aquaculture Society, Baton Rouge, LA.182 pp.

Avnimelech ,Y., Mokady, S.,& Schoroder, G.L., (1989). Circulated ponds as efficient bioreactors for single cell protein production. Bamdigeh. 41(2),58–66.

Badiola, M., Mendiola, D., & Bostock, J. (2012). Recirculating Aquaculture Systems (RAS) analysis: main issues on management and future challenges. Aquacultural Engineering 51: 26-35. https://doi.org/10.1016/j.aquaeng.2012.07.004

Ballester, E.L.C., Abreu, P.C., Cavalli, R.O., Emerenciano, M., Abreu, L., & Wasielesky, W. (2010). Effect of practical diets with different protein levels on the performance of Farfantepenaeus paulensis juveniles nursed in a zero exchange suspended microbial flocs intensive system. Aquaculture Nutrition,16(2),163-172. https://doi.org/10.1111/j.1365-2095.2009.00648.x

Browdy, C.L., Bratvold, D., Stokes, A.D., & Mcintosh, R.P. (2001). Perspectives on the application of closed shrimp culture systems. In: Jory ED, Browdy CL, editors. The new Wave, Proceedings of the Special Session on Sustainable Shrimp Culture, The World Aquaculture Society, Baton Rouge, LA, USA. pp.20–34.

Burford, M.A., Thompson, P. J., McIntosh, R. P., Bauman, R .H., & Pearson, D.C. (2003). Nutrient and microbial dynamics in high-intensity, zero-exchange shrimp ponds in Belize. Aquaculture, 219(1-4), 393-411.

Burford, M.A., Thompson, P.J., McIntosh, R.P., Bauman, R.H., &Pearson, D.C. (2004). The contribution of flocculated material to shrimp (Litopenaeus vannamei) nutrition in a highintensity, zero-exchange system. Aquaculture, 232(1-4),525–537. https://doi.org/10.1016/S0044-8486(03)00541-6

Crab, R,, Kochva, M., Verstraete, W., & Avnimelech, Y. (2009). Bio-flocs technology application in over-wintering of tilapia. Aquaculture Engineering, 40(3),105-112. https://doi.org/10.1016/j.aquaeng.2008.12.004

Crab, R., Chielens, B., Wille, M., Bossier, P., & Verstraete, W. (2009). The effect of different carbon sources on the nutritional value of bioflocs, a feed for Macrobrachium rosenbergii postlarvae. Aquaculture Research, 41(4),559–567.https://doi.org/10.1111/j.1365-2109.2009.02353.x

Crab, R., Defoirdt, T., Bossier, P., &Verstraete, W. (2012). Biofloc technology in aquaculture: beneficial effects and future challenges. Aquaculture, 356,351–356.https://doi.org/10.1016/j.aquaculture.2012.04.046

Defoirdt, T., Boon, N., Bossier, P. & Verstraete, W. (2004). Disruption of bacterial quorum sensing: an unexplored strategy to fight infections in aquaculture. Aquaculture, 240(1-4), pp.69-88.

De Schryver, P., Crab, R., Defoirdt, T., Boon, N., &Verstraete,W. (2008). The basics of bio-flocs technology: the added value for aquaculture. Aquaculture, 277(3-4),125-137. https://doi.org/10.1016/j.aquaculture.2008.02.019

Emerenciano, M. G. C., Martínez-Córdova, L. R., Martínez-Porchas, M., & Miranda-Baeza, A. (2017). Biofloc technology (BFT): a tool for water quality management in aquaculture. Water quality, 5, 92-109.

Hargreaves, J.A. (2006) Photosynthetic suspended-growth systems in aquaculture. Aquaculture Engineering, 34(3), 344-363.
https://doi.org/10.1016/j.aquaeng.2005.08.009

Jiang, S. (2010). Aquaculture, capture fisheries, and wild fish stocks. Resource and Energy Economics, 32(1),65 77.

Ju, Z.Y., Forster, I., Conquest, L., Dominy, W., Kuo, W.C. & Horgen, F.D (2008). Determination of microbial community structures of shrimp floc cultures by biomarkers and analysis of floc amino acid profiles. Aquaculture Research, 39(2), 118–133.https://doi.org/10.1111/j.1365-2109.2007.01856.x

Karimanzira, D., Keesman, K., Kloas, W., Baganz, D., & Rauschenbach, T. (2017). Efficient and economical way of operating a recirculation aquaculture system in an aquaponics farm. Aquaculture Economics & Management, 21(4),470–486. https://doi.org/10.1080/13657305.2016.1259368

Krummenauer, D., Peixoto, S., Cavalli, R.O., Poersch, L.H., & Wasielesky, W, (2011). Superintensive culture of white shrimp, Litopenaeus vannamei, in a biofloc technology system in southern Brazil at different stocking densities. Journal of World Aquaculture Society, 42(5),726–733

Kuhn, D.D., Boardman, G.D., Lawrence, A.L., Marsh, L., & Flick, G.J. (2009). Microbial floc meal as a replacement ingredient for fish meal and soybean protein in shrimp feed. Aquaculture, 296,51–57.

Kumar, M., & Lin, J.G. (2010). Co-existence of anammox and deni- trification for simultaneous nitrogen and carbon removal— strategies and issues. Journal of Hazardous Materials, 178(1-3),1–9.https://doi.org/10.1016/j.jhazmat.2010.01.077

Manan,H., Julia, H., Moh, Z., . Kasan, N.A., Suratman, S., & Ikhwanuddin M. (2016). Study on carbon sinks by classified biofloc phytoplankton from marine shrimp pond water, AACL Bioflux 9 4.http://hdl.handle.net/123456789/5751

Nunes, A.J.P., Sa, M.V.C., Andriola-Neto, F.F. & Lemos, D (2006). Behavioral response to selected feed attractants and stimulants in Pacific white shrimp, Litopenaeus vannamei. Aquaculture, 260(1-4), 244–254.

Ogello, E.O., Outa, N.O., Obiero, K.O., Kyule, D.N. & Munguti, J.M (2021). The prospects of biofloc technology (BFT) for sustainable aquaculture development. Scientific African, 14, p.e01053. https://doi.org/10.1016/j.sciaf.2021.e01053

Otoshi, C. A., Tang, L. R., Moss, D. R., Arce, S. M., Holl, C. M., & Moss, S. M. (2009). Performance of Pacific white shrimp, Penaeus (Litopenaeus) vannamei, cultured in biosecure, super-intensive, recirculating aquaculture systems. The Rising Tide–Proceedings of the Special Session on Sustainable Shrimp Farming. The World Aquaculture Society, Baton Rouge, Louisiana, USA, 137, 244-252.

Perez-Fuentes, J.A., Perez-Rostro, C.I., & Hernandez-Vergara, M.P. (2013). Pond-reared Malaysian prawn Macrobrachium rosenbergii with the biofloc system. Aquaculture, 400,105–110.

Piedrahita, R.H. (2003). Reducing the potential environmental impact of tank aquaculture effluents through intensification and recirculation. Aquaculture, 226,35–44.https://doi.org/10.1016/S0044-8486(03)00465-4

Péron, G., Mittaine, J. F., & Le Gallic, B. (2010). Where do fishmeal and fish oil products come from? An analysis of the conversion ratios in the global fishmeal industry. Marine policy, 34(4), 815-820.

Rahman, M.M., Nagelkerke, L.A., Verdegem, M.C., Wahab, M.A., & Ver- reth, J.A. (2008). Relationships among water quality, food resources, fish diet and fish growth in polyculture ponds: a multivariate approach. Aquaculture, 275(1-4),108–115.https://doi.org/10.1016/j.aquaculture.2008.01.027

Rakocy, J.E., Bailey, D.S., & Thoman, E.S., Shultz, R.C. (2004). Intensive tank culture of tilapia with a suspended, bacterial based treatment process: new dimensions in farmed tilapia. In: Bolivar R, Mair G, Fitzsimmons K, editors. Proceedings of the Sixth International Symposium on Tilapia in Aquaculture, pp. 584–596.

Ray, A. (2012). Biofloc technology for super-intensive shrimp culture. In: Biofloc Technology - a practical guide book, 2nd ed., The World Aquaculture Society, Baton Rouge, Louisiana, USA. pp. 167-188.

Sharifinia, M., Afshari Bahmanbeigloo, Z., Smith, W.O Jr, Yap, C.K., & Keshavarzifard, M. (2019). Prevention is better than cure: Persian Gulf biodiversity vulnerability to the impacts of desalination plants. Global Change Biology, 25,4022–4033.https://doi.org/10.1111/gcb.14808

Sharifinia, M., Taherizadeh, M., Namin, J.I., & Kamrani, E. (2018). Ecological risk assessment of trace metals in the surface sediments of the Persian Gulf and Gulf of Oman: evidence from subtropical estuaries of the Iranian coastal waters. Chemosphere, 191: 485–493. https://doi.org/10.1016/j.chemosphere.2017.10.077

Silva, K.R.D., Wasielesky, W. & Abreu, P.C. (2013) N and P dynamics in the biofloc production of the pacific white shrimp, Litopenaeus vannamei. Journal of the World Aquaculture Society, 44(1), 30–41.https://doi.org/10.1111/jwas.12009

Wasielesky, W.Jr., Atwood, H., Stokes, A., & Browdy, C.L. (2006). Effect of natural production in a zero exchange suspended microbial floc based super-intensive culture system for white shrimp Litopenaeus vannamei. Aquaculture, 258(1-4),396–403. https://doi.org/10.1016/j.aquaculture.2006.04.030

Y.Avnimelech (2012). Biofloc Technology- A practical guide book, The World Aquaculture Society, 2nd ed., Baton Rouge, Louisiana, United States.

Y. Avnimelech. (1999). Carbon and nitrogen ratio as a control element in Aquaculture systems, Aquaculture, 176(3-4) , 227–235.

6

Application of Medicinal Plant in Fish Feed and Its Challenges

Monika[1*], Kamin Alexander[1] and Durgesh Kumar Verma[2]

[1]Sam Higginbottom University of Agriculture Technology and Sciences, Department of Biological Sciences, Prayagraj-211007, Uttar Pradesh, India

[2]ICAR-Central Inland Fisheries Research Institute, Barrackpore, Best Bengal Regional Centre, Prayagraj-211002, Uttar Pradesh, India

Abstract

The integration of medicinal plants into fish feed presents a promising strategy for addressing health challenges in aquaculture while promoting sustainability. Medicinal plants offer a diverse array of bioactive compounds with antimicrobial, antioxidant, immunostimulatory, and anti-inflammatory properties, which can enhance fish health and disease resistance. Supported by both traditional knowledge and scientific research, medicinal plants have demonstrated efficacy in improving immune parameters, reducing mortality rates, and enhancing overall fish fitness. Moreover, their use aligns with the growing demand for sustainable aquaculture practices by offering a natural and eco-friendly alternative to synthetic medications. However, challenges such as maintaining nutritional balance, addressing toxicity concerns, ensuring digestibility and palatability, and navigating regulatory requirements accompany the utilization of medicinal plants in fish feed. Mitigation strategies including nutritional analysis, toxicity screening, processing techniques, palatability enhancement, quality assurance, research and development, regulatory compliance, and education are essential to overcome these challenges effectively. The integration of medicinal plants into fish feed signifies a paradigm shift towards sustainable and holistic aquaculture practices. Through continued research, innovation, and collaboration, the full potential of medicinal plants in optimizing fish health, welfare, and productivity can be realized, contributing to the resilience and sustainability of the aquaculture industry.

Keyword: *Medicinal Plant, Fish Feed, Aquaculture, Bioactive Compounds.*

1. Introduction

For millennia, civilizations worldwide have relied on medicinal plants for healthcare, with their usage documented as far back as ancient Sumeria. Today, traditional medicinal plants remain vital in many regions, especially in developing countries. Ethnobotanical studies have identified numerous bioactive plants, sparking extensive research into their biological activities and chemical compositions. Interest in medicinal plants has surged due to the drawbacks and costs of prescription drugs, leading to increased exploration of plant-based alternatives. Plants offer diverse chemical compositions with varied biological activities, making them suitable for treating complex diseases and posing little risk of antibiotic resistance development. Aquaculture, the fastest-growing sector in animal food production, faces challenges such as overcrowding and poor water quality, leading to disease outbreaks. While veterinary drugs are commonly used, their intensive use poses environmental and health risks, including antibiotic resistance and accumulation in animal tissues. Vaccines are effective but expensive and often limited in scope. Given the drawbacks of synthetic drugs, there's a growing need for alternative disease management strategies in aquaculture. Medicinal plants offer a promising solution, with reported bioactivities such as stress reduction, immune stimulation, and antiparasitic effects. By enhancing fish immunity and fitness, medicinal plants provide a cheaper and more sustainable alternative to chemotherapy in aquaculture *(Cabello et al., 2006; Romero-Ormazabal et al., 2012).*

The application of medicinal plants in fish feed represents a promising avenue in aquaculture for promoting fish health, improving disease resistance, and enhancing overall production sustainability. As concerns about the use of synthetic medications and their potential environmental impacts grow, there's increasing interest in natural alternatives, and medicinal plants offer a compelling solution.

In this context, the introduction of medicinal plants into fish feed serves as a proactive approach to address various health challenges commonly encountered in aquaculture. These plants contain a myriad of bioactive compounds with antimicrobial, antioxidant, immunostimulatory, anti-inflammatory, and growth-promoting properties, among others. By harnessing the therapeutic potential of medicinal plants, aquaculture practitioners can mitigate disease risks, reduce reliance on synthetic drugs, and foster a more environmentally sustainable production system. In this introduction, we'll explore the diverse applications of medicinal plants in fish feed, highlighting their potential benefits and implications for aquaculture practices. From enhancing immune function to reducing stress and improving growth performance, the incorporation of medicinal plants offers multifaceted advantages that align with the evolving needs of the aquaculture industry. effects *(Reverter et al., 2014)*. Moreover, by embracing natural alternatives, aquaculture stakeholders can contribute to the promotion of ecological balance and the preservation of aquatic ecosystems.

As we delve deeper into the topic, it becomes evident that the utilization of medicinal plants in fish feed represents not only a practical solution to health management but also a paradigm shift towards more sustainable and holistic aquaculture practices. Through research, innovation, and collaboration, we can unlock the full potential of medicinal plants in optimizing fish health, welfare, and productivity while fostering resilience in the face of emerging challenges.

2. Medicinal Plants

Traditionally, medicinal plants have been relied upon for preventing and treating diseases, with various parts of plants utilized for their active compounds. These plants offer a wide range of biological activities, including immune modulation, growth promotion, antioxidant enhancement, stress reduction, digestion stimulation, and gastroprotection. Importantly, medicinal plant extracts exhibit diverse molecular compositions, making them less prone to inducing drug resistance and environmentally friendlier compared to synthetic drugs. (Caruso *et al.,* 2013). In fish farming, medicinal plants are typically applied through oral administration, immersion baths, or injection. Oral administration is the most common method due to its economic feasibility and low stress on fish, although it may have slower absorption rates. Immersion baths are suitable for smaller fish populations but can be costly due to the need for large solution volumes. Intraperitoneal injection is effective for delivering precise doses quickly into the bloodstream but is labor-intensive and unsuitable for fry and fingerlings. Combining medicinal plant extracts has been shown to provide enhanced benefits in treating fish diseases. (Citarasu, 2010; Chakraborty and Hancz, 2011).

2.1 Biological Activity of Medicinal Plants in Aquaculture

Numerous studies demonstrate the efficacy of medicinal plants in promoting appetite, weight gain, and nutrient utilization in cultured fish. For instance, grouper fed with a diet containing a mixture of methanolic herb extracts displayed significantly higher weight gain compared to control fish. Additionally, plant extracts have been shown to enhance immune parameters in fish, leading to increased survivability when challenged with pathogens like Aeromonas hydrophila. (Putra *et al.*, 2013; Talpur *et al.*, 2013). Fish treated with medicinal plants also exhibit improved hematological parameters, indicating better fitness. Furthermore, both in vitro and in vivo studies highlight the potential of medicinal plants in combating marine pathogens such as bacteria, viruses, fungi, and ectoparasites. These plants demonstrate antibacterial, antiviral, antifungal, and antiparasitic activities, contributing to reduced mortality rates and enhanced immunity in cultured fish. (Wu *et al.,* 2011; Ji *et al.,* 2012).

2.2 Application of Medicinal Plants in Aquaculture

The administration of medicinal plants in aquaculture can involve various methods, including oral administration, injection, and immersion baths. However, oral administration is often considered the most suitable method due to its practicality and effectiveness in inducing physiological changes in fish. Different parts of plants and various solvents can be utilized in preparation, affecting the bioactivity and chemical composition of the extracts. It's crucial to understand these factors to ensure the desired effects and avoid toxic outcomes. Additionally, the duration and dosage of treatment play pivotal roles in achieving optimal results, as inappropriate doses can lead to adverse effects. Identifying natural plant products and understanding their modes of action are essential for maximizing their efficacy and ensuring safe usage in aquaculture. More research is needed to explore the bioactivity, compounds, and modes of action of medicinal plants, as well as their effects on fish physiology and appropriate dosing. Structured research plans involving in vitro and in vivo tests are recommended to evaluate the efficacy and safety of medicinal plants in aquaculture, considering factors such as fish immunity, haematological indicators, and gene expression analysis.

2.3 Advantage of Medicinal Plant in Fish Feeding.

Interest in using medicinal plants in aquaculture has grown significantly, but rural fish farmers have long utilized them. Research in West Java found that 46% of surveyed fish farmers used plants, often traditional remedies also used in human medicine. These plants were typically added directly to rearing water to enhance water quality, reduce stress, boost fish resistance to pathogens, and treat diseases. Farmers relied on personal experience and ethnic traditions to determine plant use and dosage. Scientific studies support the use of medicinal plants in aquaculture, with reported benefits including stress reduction, growth promotion, appetite stimulation, immune stimulation, and pathogen control. Plants contain various active compounds such as alkaloids, terpenoids, tannins, saponins, and flavonoids, contributing to their efficacy in promoting fish and shrimp health in aquaculture settings. Using medicinal plants in fish feed can offer several benefits, including enhancing fish health, improving disease resistance, and potentially reducing the need for synthetic medications. Here's how medicinal plants can be applied in fish feed:

2.3.1 Antimicrobial Properties

Many medicinal plants possess antimicrobial properties that can help control bacterial, fungal, and parasitic infections in fish. Incorporating these plants into fish feed can reduce the prevalence of diseases and improve overall fish health.

2.3.2 Antioxidant Effects

Certain medicinal plants are rich in antioxidants, such as flavonoids and polyphenols, which can protect fish cells from oxidative damage caused by free radicals. Including these plants in fish feed can enhance the antioxidant defense system of the fish, promoting better health and growth.

2.3.3 Immunostimulatory Effects

Some medicinal plants have immunostimulatory effects, meaning they can boost the immune system of fish. By incorporating these plants into fish feed, it's possible to improve the fish's ability to fight off pathogens and reduce the occurrence of diseases. (Ghosh *et al.*, 2018).

2.3.4 Anti- inflammatory Effects

Chronic inflammation can negatively impact fish health and growth. Medicinal plants with anti-inflammatory properties can help alleviate inflammation in fish tissues, promoting faster recovery from injuries and diseases. Including these plants in fish feed can contribute to overall better health.

2.3.5 Natural Growth Promoters

Certain medicinal plants contain bioactive compounds that can act as natural growth promoters for fish. These compounds may enhance nutrient absorption, improve digestion, and stimulate metabolic processes, leading to better growth rates and feed conversion efficiency.

2.3.6 Stress Reduction

Medicinal plants with adaptogenic properties can help fish cope with various stressors, such as environmental changes, handling, or transportation. By reducing stress levels, these plants contribute to overall fish welfare and improve resilience to disease.

2.3.7 Environmental Sustainability

Using medicinal plants in fish feed aligns with the growing demand for sustainable aquaculture practices. Instead of relying solely on synthetic medications, incorporating medicinal plants into fish feed offers a more natural and eco-friendly approach to promoting fish health and well-being.

However, it's essential to note that the efficacy of medicinal plants in fish feed can vary depending on factors such as plant species, dosage, preparation methods, and the specific requirements of the fish species. Therefore, it's crucial to conduct research and experimentation to determine the most effective formulations for achieving the desired outcomes in aquaculture operations.

2.3.8 Growth promoters

Numerous studies have demonstrated that utilizing medicinal plants in various forms, such as crude extracts, semi-extracts, or pure extracts, can stimulate appetite, promote weight gain (WG), and enhance the specific growth rate (SGR) of fish species. This growth-promoting effect is generally attributed to the ability of medicinal plants to increase the secretion of digestive enzymes, leading to improved nutrient absorption and utilization, ultimately resulting in growth promotion and increased survival rates in fish. crude extracts of *Allium sativum* (garlic) increased WG, SGR, and feed intake in Nile tilapia *(Oreochromis niloticus)*. Similarly, the crude powder of caraway exhibited significant growth-promoting effects in Nile tilapia, Additionally, extracts of *Zingiber officinalis* (ginger), *Cynodon dactylon, Tridax procumbens, Piper longum, and Phyllanthus niruri* were found to enhance the retention rate in *Epinephalus tauvina,* further supporting the growth-promoting properties of medicinal plants. (Awad and Awaad, 2017).

2.3.8.1 Immunostimulants

With the rapid expansion of the aquaculture industry, researchers have focused heavily on studying the immune systems of fish. Medicinal plants containing bioactive compounds have been investigated for their potential to act as immunostimulants, enhancing both specific and non-specific immune responses in fish. By bolstering the immune system, medicinal plants can improve fish's ability to defend against pathogens, ultimately reducing losses in fish production before the onset of disease. Studies have shown that the application of medicinal plants can enhance various immunological parameters in fish, including lysozyme activity, phagocytic activity, respiratory burst, complement activities, peroxidase, and anti-protease activities. However, the effectiveness of medicinal plants depends on factors such as dosage, the specific types of plants used, the active compounds present, and the size of the fish. In summary, medicinal plants have the potential to improve fish health by boosting their immune responses, thereby increasing their resistance to pathogens and reducing production losses in aquaculture. (Ghosh *et al.*, 2018).

2.3.8.2 Lysozyme activity

Lysozyme is a potent antimicrobial and antioxidant component that acts by lysing peptidoglycan in bacterial cell walls, thereby controlling microbial growth. It plays a crucial role in developing the non-specific or innate immunity of fish and shellfish by stimulating and activating complement components, neutrophils, and macrophages, ultimately leading to the phagocytosis of bacteria, parasites, and viruses as an opsonin. Lysozyme is particularly important during the early developmental stages of fish when specific immunity has not yet fully developed, helping larvae fight against microorganisms. The upregulation of serum lysozyme levels indicates the advancement of different humoral immunity mechanisms, which protect fish from various infections. (Yang *et al.*, 2015). Several medicinal plants have been reported to enhance lysozyme activity in fishes (Table 1).

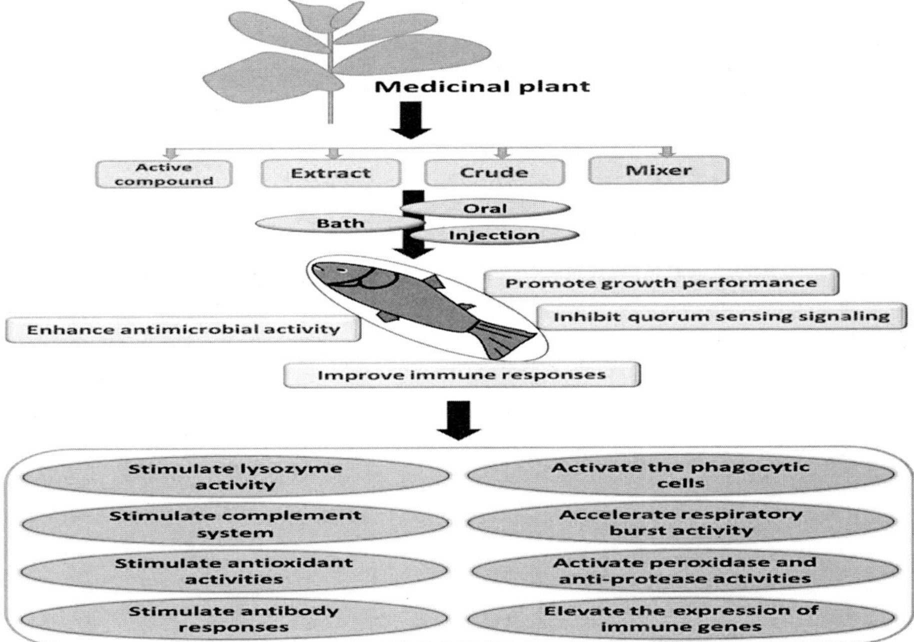

Fig. 1. Role of medicinal plants on immune status in fish

2.3.8.3 Phagocytic and respiratory burst activity

Phagocytosis serves as a primitive defense mechanism against invading pathogens, involving the internalization, killing, and digestion of these pathogens. During phagocytosis, phagocytes produce reactive oxygen species through oxidative burst activity, which is crucial for killing bacteria and other pathogens. This activity signifies the active status of macrophages and neutrophils within the cell. For vertebrates, phagocytosis is a fundamental defense mechanism, accompanied by the release of highly antimicrobial reactive oxygen species such as superoxide anion (O_2^-), hydrogen peroxide (H_2O_2), and hydroxyl radical (OH). Antioxidant enzymes like superoxide dismutase (SOD), catalase, and glutathione peroxidase help protect cells from oxidative damage. (Carbone & Faggio, 2016). Several studies have shown that dietary supplementation with medicinal plants can enhance phagocytic and respiratory burst activity in fish, leading to reduced susceptibility to pathogens (Table 1).

2.3.9 Peroxidase and anti-protease activity

Peroxidases are enzymes within a large group that play a clinically significant role as microbicidal agents. They utilize oxidative radicals to kill pathogens and contribute to the ionic stability of immune cells. During the respiratory burst, peroxidases are primarily released by neutrophils. Additionally, the anti-protease activities of serum are responsible for preventing proteolytic damage by pathogens. Several medicinal plants have been reported to elevate peroxidase activity in fish, leading to enhanced survival against various pathogens (Table 1).

2.3.10 Enhance immune gene profile

Immune-related genes such as interleukin 1 (IL-1β), interleukin 8 (IL-8), tumor necrosis factor-alpha (TNF-α), heat shock proteins (Hsp), β-defensins, and transforming growth factor-beta (TGF-β) play critical roles in defense mechanisms and can also have growth-promoting effects. IL-1β, IL-8, and TNF-α are early inflammatory cytokines secreted during inflammation. They induce host immune responses to pathogenic infections and are crucial for combating microbial invasions that can lead to tissue and organ damage. Numerous studies have shown that medicinal plants can modulate the expression of immune-related genes in various fish species. For example, supplementation with Ficus carica polysaccharide in grass carp diets led to up-regulation of IL-1β and TNF-α expression. In zebrafish, dietary supplementation of apple cider vinegar (ACV) induced the expression of lysozyme and interleukin 8 (IL-8). However, no significant changes were observed in IL-1β and TNF-α expression levels. The effect of medicinal plants on immune gene expression can be dose and tissue-dependent. For instance, olive leaf extract exhibited a dose and tissue-dependent modulation of IL-1β, IL-8, and TNF-α expression in rainbow trout. Similarly, caper supplementation in rainbow trout diets led to a significant increase in cytokine gene expression, including IL-1β, IL-8, IL-10, TGF-β, and IL-12p40, while TNF-α was downregulated. The administration of flavonoids from Heliotropium huascoense in Atlantic salmon induced the expression of TNF-α, IL-1, interferon-alpha (IFN-α), interferon-gamma (IFN-γ), and TGF-β1 in the head kidney. However, flavonoids from *Heliotropium sinuatum* decreased the transcriptional expression of TNF-α, IL-1, and IL-12. (Zou & Secombes, 2016) Overall, the effect of medicinal plants on fish immunity varies depending on factors such as fish species, immune organ, type of medicinal plants, and their concentrations.

2.3.11 Enhance antimicrobial activity

Medicinal plant extracts with antimicrobial properties offer potential treatments for diseases caused by various pathogens like bacteria, viruses, fungi, or parasites.

2.3.11.1 Antiviral activity

Lymphocystis disease virus (LDV), viral haemorrhagic septicaemia virus (VHSV), and aquabirnavirus (ABV) are significant pathogens causing economic losses to fish farmers. Extracts of *Punica granatum* increased survival rates in olive flounder infected with LDV, while *Cynodon dactylon* showed promise in treating black tiger shrimp infected with white spot syndrome virus (WSSV), resulting in no fatalities compared to controls with 100% mortality (Harikrishnan *et al.*, 2010).

2.3.11.2 Antibacterial activity

Various medicinal plants have demonstrated antibacterial properties against both gram-positive and gram-negative bacteria. For instance, cinnamon exhibited antagonistic effects against *Aeromonas hydrophila* in Nile tilapia. *Punica granatum* extracts were effective against a range of bacteria including methicillin-sensitive *Staphylococcus aureus* (MSSA), methicillin-resistant *S. aureus* (MRSA), *Escherichia coli O157:H7, Salmonella typhi,* and certain streptococci strains. (Van Hai, 2015).

2.3.11.3 Antifungal/antiparasitic activity

Prominent herbs have demonstrated antifungal or anti-parasitic effectiveness in aquaculture. Ethanolic extracts of *Piper guineense* (fruits) and *Xylopia aethiopica* (seeds) exhibited antifungal activity against *Candida albicans. Datura metal Linn* (Thorn-apple) showed significant efficacy against fungal fish pathogens, particularly effective against *Penicillium* restrictum fungal infection. (Madhuri *et al.*, 2012).

2.3.12 Alternative to antibiotics

Medicinal plants can serve as promising antibiotics for fish species by enhancing their immune status to resist diseases. *Azadirachta indica* boosts primary and secondary antibody responses in *O. mossambicus*, potentially serving as an alternative to antibiotics against *Citrobacter freundii* infection. Similarly, *Ocimum sanctum* enhances antibody production and disease resistance in *Nile tilapia* against *A. hydrophila* infection. *Sargassum duplicatum* and *Sargassum wightii* offer alternatives to antibiotics for preventing white spot *syndrome virus* disease in black tiger prawns. Additionally, a combination of four herbs could be used as an alternative to antibiotics for treating enteritis in grass carp. (Van Hai, 2015; Thanigaivel *et al.*, 2015).

2.3.13 As quorum sensing inhibitor

Quorum sensing (QS) or *Quorum signaling* (QS) is a communication system among bacteria that regulates various processes including adhesion, growth, virulence, antibiotic resistance, and biofilm formation. Due to increasing bacterial drug resistance, there's growing interest in inhibiting QS signaling. Medicinal plants produce secondary metabolites such as flavonoids, phenols, saponins, terpenoids, alkaloids, etc., which significantly affect the QS system of bacteria like *Escherichia coli*, *Staphylococcus aureus*, and *Chromobacterium violaceum*. Essential oils from aromatic plants also demonstrate activity against bacterial strains like *Salmonella, Listeria, Pseudomonas, Staphylococcus,* and *Lactobacillus* sp. Moreover, extracts from *Terminalia bellerica hinder* the QS communication system of *Pseudomonas aeruginosa* (Ganesh & Rai, 2018).

Table 1. List of medicinal plants and their extract used as immunostimulant

Medicinal plants	Fish species	Dose	Impacts
Ethanolic extracts of propolis	*Oreochromis mossambicus*	2, 4 g kg-1	Increase lysozyme and myeloperoxidase activity, total protein and disease resistance against *Streptococcus iniae.*
Essential oil from orange peel *(Citrus sinensis)*		1, 3, and 5 g kg-1	
Mucuna pruriens and *Cucurbita mixta* seed meal		2, 4 and 6 g kg-1	Enchance lysozyme, phagocytic, respiratory burst, complement activity, weight gain, feed efficiency ratio and survival against *Aeromonas hydrophila*
Cinnamon nanoparticles (CNP)	*O. niloticus*	3 g CNP kg-1	Enhance antioxidant and digestive enzymes, activities, high survival against *A. hydrophila*
Astragalus polysaccharides		1500 mg kg-1 diet	Improve growth performance, increase phagocytic, respiratory burst, lysozyme, bactericidal and amylase activity
Propolis and *Aloe barbadensis* (1:1)		0.5, 1.0 and 2%	Elevate serum lysozyme, phagocytic and antimicrobial activity.
Green tea and Chinese herbal mixtures		0.5 – 2%	Increase lysozyme, peroxidase, superoxide dismutase activity and reduce mortality against *A. hydrophila*
Peppermint (Metha piperita) plant extract	*Onchorhynchu-s mykiss*	0, 1, 2, 3%	Increase respiratory burst acitivity, total protein and protection against *Yersinia ruckeri*
Ajwain *(Trachyspermum ammi)* and marjoram (*Origanum* sp.) extract		1-2%	Improve growth performance and lysozyme activity
Lentinula edodes mushroom extract		1-2%	Decrease mortality rate against *Lactococcus garvieae* and increase *lysozyme* activity
Ficuscarica polysaccharide	*Ctenopharyng-odon idella*	1.0%	Increase lysozyme and bactericidal activity

Medicinal plants	**Fish species**	**Dose**	**Impacts**
Dill *(Anethum graveolens)* and garden cress *(Lepidium sativum)*	*Cyprinus carpio*	1-2 g kg-1	Increase lysozyme, myeloperoxidae and survival rate against *A. hydrophila* and *Edwardsiella tarda*
Ginger *(Zingiber officinale)* and Garlic *(Allium sativum)*	*Lates calcarifer*	5, 10 g kg-1	Increase phagocytic, respiratory burst, bactericidal and antiprotease activity, reduce susceptibility to *Vibrio harveyi*
Peppermint powder			Increase phagocytic, respiratory burst, bactericidal and antiprotease activity, elevate serum protein and globulin level, reduce susceptibility to *V. harveyi*
Peppermint plant extract	*Rutilus frisii kutum and Salmo trutta*	1%, 2%, and 3% respiratory only	Increase blood leukocyte number, lysozyme, antimicrobial and respiratory burst activity
Clove basil *(Ocimum gratissimum)* leaf extract	*Clarias gariepinus*	10, 15 g kg-1	Increase intestinal villi length, absorption area, reduce cholesterol and glucose level, increase protein, antioxidant and survival against *Listeria monocytogenes*
Plantagoasiatica and *Houttuynia cordata*	*Rachycentron canadum*	10, 20 g kg-1	Induce phagocytosis, respiratory burst, lysozyme activity.
Pontogammarus maeoticus extract	*Rutilus caspicus*	2%	Improve growth performance, feed intake salinity stress resistance, complement and lysozyme activity
Fenugreek *(Trigonella foenum graecum)*	*Sparus aurata*	5 and 10%	Increase haemolytic complement, peroxidase, antiprotease activity, enhance cellular and humoral immune parameters.

2.4 Challenges

Feeding fish through medicinal plants presents several challenges, ranging from nutritional considerations to potential toxicity issues. While medicinal plants may offer various health benefits to fish, their incorporation into fish diets requires careful attention to ensure both the fish's well-being and the effectiveness of the medicinal properties. Here are some key challenges:

2.4.1 Nutritional Balance: One of the primary challenges is maintaining a balanced diet for the fish. While medicinal plants may contain beneficial compounds, they may lack essential nutrients required for the fish's growth and health. Therefore, formulating a diet that meets the nutritional requirements of the fish while incorporating medicinal plants is crucial.

2.4.2 Toxicity Concerns: Many medicinal plants contain bioactive compounds that, in high concentrations, can be toxic to fish. For example, certain alkaloids, glycosides, or essential oils present in medicinal plants might adversely affect fish health if not appropriately regulated. Proper dosage and understanding of potential toxicity are essential to avoid harm to the fish.

2.4.3 Digestibility: The digestibility of medicinal plants by fish can vary significantly. Some plants may contain complex compounds that are difficult for fish to digest, leading to inefficient nutrient utilization or digestive issues. Ensuring that the chosen medicinal plants are digestible by the target fish species is critical for optimizing feed efficiency and promoting growth.

2.4.4 Palatability: Medicinal plants often have distinct flavors and odors that may affect the palatability of fish feed. Fish may reject feed containing unfamiliar or unappealing plant ingredients, leading to reduced feed intake and potential growth inhibition. Masking or enhancing the palatability of medicinal plant-based feeds through appropriate processing or supplementation can help overcome this challenge.

2.4.5 Quality and Availability: Consistency in the quality and availability of medicinal plants can be challenging, particularly for small-scale fish farmers or in regions with limited access to diverse plant resources. Ensuring reliable sourcing and maintaining the quality of medicinal plants throughout storage and processing is essential for the consistency and efficacy of the fish feed.

2.4.6 Regulatory Considerations: Incorporating medicinal plants into fish feed may involve regulatory hurdles related to safety, labeling, and approval processes. Compliance with regulations governing the use of medicinal plants in animal feed is necessary to ensure legal and ethical practices.

Addressing these challenges requires interdisciplinary collaboration among nutritionists, veterinarians, agronomists, and aqua-culturists to develop safe, effective, and sustainable fish feeds incorporating medicinal plants. Through proper research, formulation, and monitoring, these challenges can be overcome, leading to the development of innovative and beneficial feeding practices for aquaculture.

Table 2. List of some Plants, algae and mushrooms used or studied for potential application in aquaculture.

Order	Family	Species	Activity	Part Used
Apiales	Apiaceae	*Angelica Membranaceus*	Immunostimulatory	Root
Apiales	Apiaceae	*Angelica sinensis*	Immunostimulatory	Root
Apiales	Araliaceae	*Kalopanax pictus*	Immunostimulatory	Shoot
Apiales	Araliaceae	*Panax ginseng*	Immunostimulatory	Root
Asterales	Asteraceae	*Siegesbeckia glabrescens*	Immunostimulatory	Whole plant
Cyatheales	Cyatheaceae	*Cyathea kanehirae*	Immunostimulatory	Leaves
Gentianales	Apocynaceae	*Anathoda vasica*	Immunostimulatory	Leaves
Gentianales	Apocynaceae	*Calotropis gigantea*	Immunostimulatory	Leaves
Lamiales	Plantaginaceae	*Picrorhiza kurrooa*	Immunostimulatory	Leaves
Piperales	Piperaceae	*Piper nigrum*	Immunostimulatory	Root
Rosales	Moraceae	*Ficus benghalensis*	Immunostimulatory	Fruits
Apiales	Umberlliferae	*Carum carvi*	Growth promotor	Leaves
Asparagales	Amaryllidaceae	*Allium tuberosum*	Growth promotor	Leaves
Asterales	Asteraceae	*Artemisia*	Growth promotor	Whole Plant
Asterales	Asteraceae	*Artemisia cina*	Growth promotor	Whole Plant
Caryophyllales	Amaranthaceae	*Alteranthera sessilis*	Growth promotor	Leave, Shoot, Fruit
Fabales	Fabaceae	*Psoralea corylifolia*	Growth promotor	Leave
Magnoliales	Myristicaceae	*Myristica fragans*	Growth promotor	Leaves
Malpighiales	Euphorbiaceae	*Tetracarpidium*	Growth promotor	Fruits
Sapindales	Rutaceae	*Zanthoxylum*	Growth promotor	Leaves
Apiales	Apiaceae	*Bupleurum chinense*	Anti parasitic	Whole Plant
Apiales	Apiaceaei	*Radix peucedani*	Anti parasitic	Whole Plant
Apiales	Araliaceae	*Kalopanax pictus*	Anti parasitic	Shoot

Caryophyllales	Polygonaceae	*Rumex obtusifolius*	Anti parasitic	Leaves
Fabaceae	Leucaena glauca	*Leucaena glauca*	Anti parasitic	Whole Plant
Lamiales	Lamiaceae	*Origanum*	Anti parasitic	Fruits
Malpighiales	Euphorbiaceae	*Euphorbia*	Anti parasitic	Whole Plant
Piperales	Piperaceae	*Piper guineense*	Anti parasitic	Shoot
Caryophyllales	Polygonaceae	*Polyonum*	Anti-fungal	Branch
Lamiales	Lamiaceae	*Origanum onites*	Anti-fungal	Shoot
Lamiales	Lamiaceae	*Thymbra spicata*	Anti-fungal	Leaves
Malpighiales	Euphorbiaceae	*Euphorbia*	Anti-fungal	Whole Branch
Malpighiales	Euphorbiaceae	*Euphorbia*	Anti-fungal	Root
Alismatales	Araceae	*Colocasia esculenta*	Anti-bacterial	Root
Apiales	Apiaceae	*Cnidium officinale*	Anti-bacterial	Leave, Fruit
Caryophyllales	Portulacaceae	*Portulaca oleracea*	Anti-bacterial	Flower
Fabales	Fabaceae	*Astragalus*	Anti-bacterial	Leaves
Gentianales	Apocynaceae	*Daemia extensa*	Anti-bacterial	Leaves
Ranuncudales	Menisperma-ceae	*Tinospora cordifolia*	Anti-bacterial	Fruits
Solanales	Convolvulaceae	*Merremia*	Anti-bacterial	Root
Acorales	Acoraceae	*Acorus calamus*	Anti-viral	Root
Asterales	Asteraceae	*Tagetes erecta*	Anti-viral	Leaves
Caryophyllales	Molluginaceae	*Glinus oppositifolius*	Anti-viral	Fruits
Gentianales	Apocynaceae	*Catharanthus roseus*	Anti-viral	Leaves
Lamiales	Acanthaceae	*Clinacanthus nutans*	Anti-viral	Whole plant
Malpighiales	Euphorbiaceae	*Acalypha australis*	Anti-viral	Leaves
Rosales	Moraceae	*Morus alba*	Anti-viral	Leaves

2.5 Strategies

Mitigating the challenges associated with feeding fish through medicinal plants requires a multifaceted approach aimed at ensuring nutritional balance, safety, and efficacy. Here are some mitigation strategies:

2.5.1 Nutritional Analysis and Supplementation: Conducting thorough nutritional analyses of both the medicinal plants and the fish's dietary requirements can help identify potential nutrient gaps. Supplementing the feed with essential nutrients lacking in medicinal plants ensures a balanced diet for the fish.

2.5.2 Toxicity Screening and Dosage Optimization: Prioritize screening medicinal plants for toxic compounds and determining safe dosage levels for incorporation into fish feed. Utilize gradual inclusion and monitoring protocols to mitigate the risk of toxicity, ensuring the safety of the fish.

2.5.3 Processing Techniques: Employing processing techniques such as grinding, drying, or extrusion can enhance the digestibility of medicinal plants, improving nutrient absorption by fish. Processing can also help mitigate palatability issues by reducing strong flavours or Odors.

2.5.4 Palatability Enhancement: Experiment with palatability enhancers such as natural flavours or attractants to mask any undesirable taste or smell of medicinal plants, encouraging fish to consume the feed willingly.

2.5.5 Quality Assurance and Traceability: Establish quality control measures throughout the supply chain to ensure the consistent quality and availability of medicinal plants used in fish feed. Implementing traceability systems helps track the origin and handling of plant materials, ensuring compliance with safety standards.

2.5.6 Research and Development: Invest in research and development initiatives to identify novel medicinal plants with desirable properties for fish health and performance. Additionally, explore innovative processing techniques and feed formulations to optimize the efficacy of medicinal plant-based feeds.

2.5.7 Regulatory Compliance: Stay informed about regulations governing the use of medicinal plants in animal feed and ensure compliance with labelling requirements and safety standards. Engage with regulatory authorities to facilitate the approval process for novel ingredients or formulations.

2.5.8 Education and Training: Provide education and training programs for fish farmers, feed manufacturers, and other stakeholders on the safe and effective use of medicinal plants in fish feed. Promote awareness of best practices for incorporating medicinal plants while minimizing risks to fish health and welfare.

By implementing these mitigation strategies, aquaculture practitioners can overcome the challenges associated with feeding fish through medicinal plants, unlocking the potential benefits of natural health-promoting compounds while ensuring the safety and well-being of the fish.

Conclusion

In conclusion, the utilization of medicinal plants in fish feed represents a promising approach to addressing various health challenges in aquaculture while promoting sustainability and holistic practices. Medicinal plants offer a plethora of bioactive compounds with diverse biological activities, including antimicrobial, antioxidant, immunostimulatory, and anti-inflammatory properties. Incorporating these plants into fish feed can enhance fish health, improve disease resistance, and potentially reduce the reliance on synthetic medications, thus mitigating environmental risks associated with intensive drug use.

The application of medicinal plants in fish feed has been supported by both traditional knowledge and scientific research, demonstrating benefits such as growth promotion, immune enhancement, stress reduction, and pathogen control. These plants have shown efficacy in enhancing immune parameters, reducing mortality rates, and improving overall fish fitness in aquaculture settings. Furthermore, medicinal plants offer a natural and eco-friendly alternative to synthetic drugs, aligning with the growing demand for sustainable aquaculture practices.

However, the utilization of medicinal plants in fish feeding comes with challenges, including maintaining nutritional balance, addressing toxicity concerns, ensuring digestibility and palatability, ensuring quality and availability, navigating regulatory requirements, and mitigating risks through research and development. By employing mitigation strategies such as nutritional analysis, toxicity screening, processing techniques, palatability enhancement, quality assurance, research and development, regulatory compliance, and education, aquaculture practitioners can overcome these challenges and harness the benefits of medicinal plants effectively.

Overall, the integration of medicinal plants into fish feed represents not only a practical solution to health management but also a paradigm shift towards sustainable and holistic aquaculture practices. Through continued research, innovation, and collaboration, the full potential of medicinal plants in optimizing fish health, welfare, and productivity can be realized, contributing to the resilience and sustainability of the aquaculture industry.

References

Awad, E. and Awaad, A. (2017). Role of medicinal plants on growth performance and immune status in fish. Fish. Shellfish. Immunol., 67: 40-54.

Cabello, F.C. (2006) Heavy use of prophylactic antibiotics in aquaculture: a growing problem for human and animal health and for the environment. EnvironmentalMicrobiology, 8, 1137–1144.

Caruso, D., Lusiastuti, A.M., Taukhid, Slembrouck, J., Komarudin, O. and Legendre, M. (2013) Traditional pharmacopeia in small scale freshwater fish farms inWest Java, Indonesia: an ethnoveterinary approach. Aquaculture, 416–417, 334–345.

Chakraborty, S.B. and Hancz, C. (2011) Application of phytochemicals as immunostimulant, antipathogenic and antistress agents in finfish culture. Reviews in Aquaculture, 3, 103–119.

Ganesh, P.S. and Rai, V.R. (2018). Attenuation of quorum-sensing-dependent virulence factors and biofilm formation by medicinal plants against antibiotic resistant Pseudomonas aeruginosa. J. Trad. Comp. Med., 8(1): 170-177.

Harikrishnan, R.; Heo, J.; Balasundaram, C.; Kim, M.C.; Kim, J.S.; Han, Y.J. and Heo, M.S. (2010). Effect of Punica granatum solvent extracts on immune system and disease resistance in Paralichthys olivaceus against lymphocystis disease virus (LDV). Fish Shellfish Immunol., 29(4): 668-673.

Ji, J., Lu, C., Kang, Y.,Wang, G.X. and Chen, P. (2012) Screening of 42 medicinal plants for in vivo anthelmintic activity against Dactylogyrus intermedius (Monogenea) in goldfish (Carassius auratus). Parasitology Research, 111, 97–104.

Madhuri, S.; Mandloi, A.K.; Govind, P. and Sahni, Y.P. (2012). Antimicrobial activity of some medicinal plants against fish pathogens. Int. Res. J. Phar., 3(4): 28-30.

Putra, A.A.S. Santoso, U., Lee, M.C. and Nan, F.H. (2013) Effects of dietary katuk leaf extract on growth performance, feeding behavior and water quality of grouper Epinephelus coioides. Aceh International Journal of Science and Technology, 2.

Reverter, M., Bontemps, N., Lecchini, D., Banaigs, B. and Sasal, P. (2014) Use of plant extracts in fish aquaculture as an alternative to chemotherapy: current status and future perspectives. Aquaculture, 433, 50–61.

Romero Ormazábal, J.M., Feijoó, C.G. and NavarreteWallace, P.A. (2012) Antibiotics in aquaculture – use, abuse and alternatives, in Health and Environment in Aquaculture (eds E.D.Carvalho, J.S. David and R.J. Silva), InTech, Croatia, p. 159.

Talpur, A.D., Ikhwanuddin, M. and Ambok Bolong, A.M. (2013) Nutritional effects of ginger (Zingiber officinale Roscoe) on immune response of Asian sea bass, Lates calcarifer (Bloch) and disease resistance against Vibrio harveyi. Aquaculture, 400–401, 46–52.

Van Hai, N. (2015). The use of medicinal plants as immunostimulants in aquaculture: A review. Aquac., 446: 88-96.

Wu, Z.F., Zhu, B.,Wang, Y., Lu, C. andWang, G.X. (2011) In vivo evaluation of anthelmintic potential of medicinal plant extracts against Dactylogyrus intermedius (Monogenea) in goldfish (Carassius auratus). Parasitology Research, 108, 1557–1563.

Yang, X.; Guo, J.L.; Ye, J.Y.; Zhang, Y.X. and Wang, W. (2015). The effects of Ficus carica polysaccharide on immune response and expression of some immune-related genes in grass carp, Ctenopharyngodon idella. Fish. Shellfish. Immunol., 42: 132-137.

Zou, J. and Secombes, C.J. (2016). The function of fish cytokines. Biol., 5: 23.

7

Application of Recent and Advanced Tools in Fish Disease Diagnosis

Arunjyoti Baruah[1*], Arya Singh[1], Bannuru Surya Chaitanya[1] Vivek Chauhan[2] and Muzammal Hoque[1]

[1]ICAR-Central Institutes of Fisheries Education, Mumbai-400061 Maharashtra, India

[2]College of Fisheries Science, CCS HAU, Hisar-12500, Haryana, India

Abstract

New techniques for fish disease diagnosis are rapidly expanding as a result of advancements in molecular biology. Conventional methods of fish pathogen identification-culture, serology, and histology-may be slowed down and rendered less sensitive by molecular techniques. Nucleic acid amplification by polymerase chain reaction (PCR), restriction enzyme digestion, probe hybridization, RFLP, AFLP and nucleotide sequencing are among the techniques of great importance. Using molecular diagnostic tools, pathogens can be identified in asymptomatic fish, potentially preventing disease spread. This chapter reviews the molecular approaches for fish pathogen identification and discusses the potential applications of these techniques.

Keywords: *Molecular diagnostics; DNA; PCR; Probes; RFLP; AFLP; Nucleotide sequencing*

1. Introduction

Aquaculture is described as an organized production of crops in an aquatic medium (FAO, 1987). Global aquaculture production has continued to expand at approximately 9% per year since the 1970s, with Asia dominating production levels, particularly for finfish (FAO, 2014). FAO data indicate that global fish production stands at 167 million tons (MT), of which 44% (73.8 MT) is contributed by the aquaculture sector (FAO, 2016). Twelve percent of the world's population depend on the fishing industry for their livelihood as and it provides 17 percent of animal protein. India is second only to China in terms of the world's output of culturable fisheries. According to reports, the aquaculture industry in India has suffered significant losses from the frequent incidence of infectious diseases of bacterial and viral origins, which has hindered sustainable development. To overcome the challenges of developing appropriate technologies for managing water quality in culture systems, developing quick diagnostic tools and ways to contain disease outbreaks, and providing disease-free or highly healthy broodstock and seeds, aquaculture requires creative biotechnological interventions "is the need of the hour" (Subasinghe *et al.*, 2003).

2. Disease problems in fish culture

The definition of the word "disease" is unclear. According to Campbell *et al.*, (1979), a disease is the culmination of all anomalous behaviors exhibited by a collection of living things in connection with a particular trait or combination of traits that sets them apart from the average for their species and puts them at a biological disadvantage. Therefore, there are two types of causes of disease: non-biological (abiotic) and biological (biotic). Kinne (1980) provided the following epidemiological description of illnesses (epizootiology for animal diseases)

- Sporadic diseases occur sporadically in a comparatively small number of individuals in a population.
- Large-scale outbreaks of infectious diseases that occur momentarily in constrained regions are known as epidemics and epizootics.
- Pandemics and panzootics are widespread outbreaks of infectious diseases that affect considerable geographic regions.
- Endemics/enzootics: These are illnesses that linger or recur as localized, low-level epidemics.

Diseases occur in aquatic environments as a result of intricate interactions between disease pathogens, host species, and the surrounding environment. Adverse environmental circumstances, pathogen virulence, and host vulnerability have a significant impact on disease outbreaks. Owing to high stocking densities, increased stress on cultured stocks, and insufficient water exchange, intensive and semi-intensive farming techniques promote the development of illness (Idowu *et al.*, 2017; Kaoud, 2015). Fish parasites can cause discomfort, decreased function, weight loss, and even death by infesting the gills, skin, stomach, or fish muscle tissue as grub-like worms.

3. Disease diagnosis

Before we define the term 'Disease', we should first understand 'What is Health?' According to the World Health Organization, Health is "a state of complete physical, mental, and social well-being, not merely the absence of disease or infirmity" (WHO, 1946). Therefore, disease' is any deviation from normal health and is defined as any harmful deviation from the normal structural or functional state of an organism, generally associated with certain signs and symptoms and differing in nature from physical injury" (EMBO reports, 2004). The fundamental principles for disease diagnosis involve three main steps: i) establishing an accurate diagnosis, ii) choosing a suitable and environmentally responsible treatment, and iii) obtaining detailed information on farm management practices to ascertain whether future outbreaks can be prevented through procedural or design changes (Bondad-Reantaso *et al.*, 2001; Bondad-Reantaso *et al.*, 2005). The three diagnostic levels for emerging and reportable fish and shellfish diseases recommended by OIE (2000) and FAO/NACA (Hamera & Bondad-Reantaso, 2001) are:

3.1 Level I: Farm/production site observations, record keeping, and health management

3.2 Level II: It includes the involvement of specialized parasitology, histopathology, bacteriology, and mycology, which require moderate capital and training investment

3.3 Level III: involves advanced diagnostic techniques, which require significant capital and training investment, isolation of causative agents such as bacteria, fungi, and viruses, biochemical characterization of pathogens, immunoassays (enzyme-linked immunosorbent assays or fluorescent, antibody technique), transmission electron microscopy, and molecular diagnostics such as polymerase chain reaction and nucleic acid assays using specific probes.

4. Recent trends in fish disease diagnosis

4.1 Molecular techniques

Developments in molecular biology have led to the rapid development of new methods for diagnosing fish diseases. Techniques of major significance include the polymerase chain reaction (PCR) amplification of nucleic acids, restriction enzyme digestion, probe hybridization, loop-mediated isothermal amplification, microarrays, and nucleotide sequencing. Many molecular techniques are potentially faster or more sensitive than methods such as culture, serology, and histology, which are traditionally used to identify diseases in fish.

4.2 Polymerase chain reaction (PCR) amplification of nucleic acids

Polymerase chain reaction (PCR) is a widely employed method for amplifying a small sample of DNA (or part of it) to billions of copies of a specific DNA sample. In 1983, polymerase chain reaction (PCR) was invented by the American biochemist Kary Mullis at Cetus Corporation, and Michael Smith was collectively honored with the Nobel Prize in Chemistry in 1993. The basic principle of PCR involves three steps: i) melting or denaturation (strand separation) of the double-stranded DNA template to form a single-strand within one–several minutes at 94-96°C. Heating at 94 °C for 30 s allowed the oligonucleotide primers to anneal to single-stranded template DNA at specific locations. If the melting temperature is too low or the time is too short, the double-stranded DNA may not denature, thereby reducing the efficiency of the reaction; ii) annealing the two primers to bind opposite DNA strands at 55°C for 30-60 sec. Recall that if the temperature is too low, the primer will bind with the templates non-specifically, and if the temperature is too high, the primer will not bind at all. iii) The primers can be extended by adding nucleotides to the 3' end of the primer sequence (adenosine pairs with thymine and guanine pairs with cytosine) using polymerase (Taq polymerase) to create two copies of the original template DNA sequence. Taq polymerase works best at a temperature range of 72 to 75°C (Leobert D. de la Peña 2002). PCR, known for its speed, sensitivity, and high specificity, is used for the rapid detection of viruses affecting fish and shrimp, including WSSV, MBV, IHHNV, Hepatopancreatic parvovirus (HPV), TSV, YHV, red sea bream iridovirus (RSIV), and infectious spleen and kidney necrosis virus (ISKNV) (Lightner, 1996; Lightner and Redman 1998: Kyung-Ho Kim *et al.*, 2024).

Table 1. Commonly used PCR techniques for fish disease diagnosis

Types of PCR	Application
Conventional (quantitative) PCR	Synthesis of particular DNA fragments with the aid of the enzyme DNA polymerase
Real-Time PCR (Quantitative PCR (qPCR))	Applied in genotyping and quantifying pathogens, analysing microRNA, detecting cancer, assessing microbial loads, and detecting GMOs by providing real-time detection of products during the exponential phase
Reverse Transcriptase PCR (RT-PCR)	Used in gene insertion, genetic disease diagnosis and cancer detection by using complementary DNA (cDNA) (converted from RNA) as a template
Amplified fragment length polymorphism (AFLP) PCR	Can swiftly produce a huge number of marker fragments for every organism without needing to know its genome sequence beforehand.
Allele-specific PCR	Likewise known as the (refractory mutation system for amplification) When detecting single gene point mutations in diseases like sickle cell anemia and thalassemia, ARMS-PCR is frequently used.
Colony PCR	To verify proper ligation and integration of transferred DNA into yeast plasmids and bacteria

Digital PCR (dPCR)	To ascertain the overall counts of bacteria, viruses, and parasites in a range of clinical specimens, primarily in situations where a well-calibrated standard is unavailable.
Fast cycling PCR	Usage in procedures calls for short cycles and aids in the prompt detection of illnesses and genetic changes.
Hot start PCR	Is a type of traditional polymerase chain reaction (PCR) that, when used at room temperature, lessens the production of primer dimers and unwanted products as a result of non-specific DNA amplification.
Multiplex PCR	Used to amplify several targets in a single PCR test run. helpful in identifying several microbes that cause the same kinds of illnesses
Nested PCR	Is a variation of PCR technique where two sets of primers are used to prevent non-specific binding, increasing the specificity of the reaction.

4.3 Restriction enzyme digestion: The enzyme that breaks down DNA into smaller pieces at or close to particular recognition sites inside molecules is called a restriction enzyme (also known as restriction endonuclease, REase, or ENase) (short, generally palindromic sequences of 4–8 bp length) (Alfred Jeltsch, Alfred Pingoud). Restriction enzymes are typically categorized into five types based on their structure: whether they can cleave their DNA substrate at the recognition site or whether they have separate recognition and cleavage sites. Type II restriction enzymes are most commonly used. More than 3000 type II restriction endonucleases (each with a unique nucleotide sequence at which it cuts a DNA molecule) have been discovered (Alfred Pingoud, Albert Jeltsch). Restriction enzymes are frequently used to create a "fingerprints" of particular DNA molecules. Every DNA molecule has unique fingerprints.

4.4 Restriction Fragment Length Polymorphism (RFLP): RFLP locates the locations of genes within a sequence and identifies changes across populations or species using variations in homologous DNA sequences, which are sometimes referred to as polymorphisms. Using one or more restriction enzymes to break the DNA test sample, the resultant fragments were separated based on size using gel electrophoresis. A given restriction endonuclease produces fragments of different lengths during digestion if the spacing between the cleavage sites differs between the two species. Using the similarity of the patterns formed, they can be used to differentiate between species and, in certain situations, even strains (Altinok and Kurt 2003).

4.5 Amplified Fragment Length Polymorphism (AFLP): In AFLP analysis, adapters are ligated to genomic restriction fragments, and adapter-specific primers are used in PCR-based amplification (Vaneechoutte, 1996).

KeyGene first developed this technology in the early 1990s and was patented in Europe in 2000 (Sahin *et al.*, 2022). AFLP technology is currently one of the most widely used genetic fingerprinting techniques worldwide. Only a small amount of purified genomic DNA was required for the AFLP analysis. This DNA is first digested by two restriction enzymes, one of which has an average cutting frequency (such as EcoRI), and the other has a higher cutting frequency (such as MseI or TaqI). The ends of the DNA fragments were then joined to linkers or adapters using restriction sites. To act as primer binding sites for the subsequent PCR amplification of the restriction fragments, linkers and surrounding restriction sites stably add nucleotides to the 3' ends of the PCR primers. Amplification was limited to restriction fragments in which the selected nucleotides and restriction site nucleotides matched. The amplified fragments were visualized by autoradiography, phosphoimaging, or other methods (Mishra *et al.*, 2020).

4.6 Random Amplified Polymorphic DNA (RAPD): A type of polymerase chain reaction (PCR), involves the amplification of random segments of DNA. In RAPD, several short, arbitrary primers (10–12 nucleotides) are employed to initiate PCR with a large genomic DNA template, with the hope that diverse fragments will undergo amplification. Random genomic DNA segments were amplified by PCR using a single primer with an arbitrary nucleotide sequence to produce DNA fragments of 10 nucleotides, which are known as RAPD markers. Using random primers, these markers can be used to distinguish between genetically dissimilar individuals, which is useful for analyzing an individual's genetic diversity. After amplification, samples were loaded into a gel (either agarose or polyacrylamide), and the different sizes created through random amplification were separated along the gel based on their size. This resulted in a unique DNA fingerprint. The primary benefit of RAPD is that, in contrast to standard PCR analysis, it does not require any particular understanding of the DNA sequence of the target organism. RAPD has also been used (Huang *et al.*, 1994). The crayfish plague fungus Astacus astaci and Aphanomyces species, which have seriously harmed fish populations in Asia's aquaculture and wild fisheries, have been studied using this methodology (Lilley *et al.*, 1997; Huang *et al.*, 1994).

4.7 DNA microarrays: A DNA microarray, often referred to as a DNA chip or biochip, is composed of microscopic DNA spots that adhere to a solid surface. Each spot contained picomoles (10–12 mol) of a unique DNA sequence or probe. Probe-target hybridization occurs when a target that contains complementary sequences is added. The relative abundance of nucleic acid sequences in a target is typically ascertained using fluorophores, silver, or chemiluminescence-labelled targets for the detection and quantification of probe-target hybridization. The fundamental concept of this technique is based on the complementary hydrogen bond binding of nucleic acid sequences between the target and probe. Microarrays provide several benefits over traditional nucleic acid hybridization on membranes, including higher sensitivity, cost-effectiveness, automation, and reduced background interference (Shalon *et al.*, 1996).

Research has demonstrated that the use of DNA microarray assay, with probes made from variable regions of the bacterial 16S rRNA gene and gryB sequences, can identify eight species of pathogenic bacteria in cultured aquatic animals, including Vibrio harveyi, V. alginolyticus, V. parahaemolyticus, V. anguillarum, V. cholerae, Nocardia seriolae, Aeromonas hydrophila, and Streptococcus iniae (Yu-Hong Shi). For the simultaneous identification of Vibrio vulnificus, Listonella anguillarum, Photobacterium damselae subsp. damselae, Aeromonas salmonicida subsp. salmonicida, and Vibrio parahaemolyticus, Santiago *et al.* developed a test using multiplex PCR and a DNA microarray (Santiago F. Gonzalez).

5. In situ Hybridization: With situ hybridization: A targeted DNA or RNA sequence is precisely located within particular tissue sections or, in the case of smaller tissues, such as plant seeds or Drosophila embryos, throughout the entire tissue (whole-mount ISH). This is achieved using labelled complementary DNA, RNA, or modified nucleic acid strands (probes). Typically, probes are 20 nucleotides or less of fluorescence-labeled DNA or RNA (Mishra *et al.*, 2020). Whole-mount in situ hybridization, double detection of RNAs and RNA plus protein, analysis using light and electron microscopes, in situ hybridization to mRNA using oligonucleotide and RNA probes (both radiolabeled and hapten-labelled), and fluorescent in situ hybridization to detect chromosomal sequences are among the in-situ hybridization techniques currently in use.

A different popular kind of in-situ hybridization is called fluorescence in situ hybridization (FISH), which is used to tag chromosomes or cells according to how complementary DNA or DNA/RNA double strands are (Mishra *et al.*, 2020). According to Chenghua *et al.* (2016), the basic idea behind FISH is based on the use of DNA strands that have been precisely modified using fluorophore-coupled nucleotides (probes) for hybridization. These probes attached to the complementary sequences inside the examined cells and tissues. Flow cytometry or fluorescence microscopy was then used to visualize the labeled cells (Altinok & Kurt, 2003).

5.1 Immunoassays: Immunoassays are essential techniques that play a crucial role in the diagnosis and management of infectious diseases by detecting the antigens of microorganisms. Antibodies and antigens react in specific ways (Mishra *et al.*, 2020). Reactions between antigens and antibodies are appropriate for recognizing one another because of their high specificity. Serology is the study of antigen-antibody responses in vitro. Antigen-antibody reactions are used in immunodiagnostic testing to identify a particular antigen or antibody linked to a pathogenic microorganism. The benefit of immunoassays is their capacity to directly determine the presence of a particular pathogen in specimens and the specific antibodies generated by the host immunological response to the organism. The antibody is the central component of the test (de la Peña and Leobert D 2001). Antibodies that develop in an animal in reaction to a single antigen are called polyclonal antibodies because they are made up of multiple distinct cell clones (heterogeneous), whereas monoclonal antibodies are uniform and originate from a single cell clone, as in the case of a plasma cell malignancy (myeloma) (de la Peña and Leobert D 2001). There are two primary techniques for identifying unknown organisms using sera containing different known antibodies: precipitation of soluble antigens and agglutination of particulate antigens. Additional immunoassays include fluorescence antibody techniques (FAT), enzyme-linked immunosorbent assay (ELISA), and western blot analysis.

5.2 Agglutination: The agglutination reaction is one of the simplest immunological tests involving particulate antigens (bacteria and red blood cells) or inert particles (latex beads) coated with an antigen. The divalent or multivalent nature of antibodies provides crosslinks with antigenically multivalent particles, forming a lattice that is visible as clumping or agglutination. According to Toranzo *et al.*, (1987), agglutination testing is a frequently used technique to detect bacterial infections in fish belonging to the genera Vibrio, Pasteurella, Aeromonas, Yersinia, Edwardsiella, and Pseudomonas.

5.3 Precipitation: Antibody molecules crosslink in varying amounts during this precipitation event, causing aggregates or precipitates to form. Three areas of antigen-antibody precipitate formation are involved in the precipitation reaction: i) the zone of equivalence, where the maximal precipitate forms because antigen and antibody have ideal proportions; ii) the zone of antibody excess, where the presence of too much antibody prevents efficient lattice formation and causes precipitation to be less than maximal; and iii) the zone of antigen excess, where all antibodies have combined with antigen and precipitation is reduced because many antigen-antibody complexes are too small to precipitate (de la Peña and Leobert D 2001). The standard precipitation techniques include immune electrophoresis, double diffusion, and single diffusion.

5.4 Fluorescent Antibody Technique (FAT): The principle of the fluorescence antibody test is that specific antibodies attach themselves to proteins or antigens. When light of a particular wavelength hits a fluorochrome, it absorbs and emits light of different wavelengths, which can be observed using a fluorescence microscope. Primary and secondary antibodies, fluorescent dyes, or fluorophores (such as fluorescein, phycoerythrin, or rhodamine) conjugated to the antibody, as well as an immunofluorescence microscope for visualization, are the primary requirements for the fluorescence antigen technique (FAT). Primary antibodies are specific antibodies that bind directly to an antigen with complementary binding sites for the antibody, whereas secondary antibodies bind to the Fc region of a primary antibody that is already bound to the specific antigen (Bikash Dwivedi 2022). FAT is divided into two categories based on its applications: i) direct fluorescent antibody (DFA) assays, which involve direct binding and illumination of a target antigen by a fluorescently tagged monoclonal antibody. The DFA test is beneficial for quickly identifying bacterial pathogens, such as Streptococcus pyogenes, Mycoplasma pneumoniae, and Legionella pneumophila (Microbiology: Canadian edition); and ii) indirect fluorescent antibody (IFA) tests, in which a fluorescein-labelled antibody conjugate is added in a second step to identify bound antibodies (de la Peña and Leobert D 2001). FAT has been widely used to detect antibodies against Aeromonas liquifaciens, Renibacterium Salmoninarumin, and Vibrio penaeicida (Lewis and Savage, 1972; Bullock *et al.*, 1980; de la Peña *et al.*, 1992) in fish, and in cases of viral infections, FAT has been developed to diagnose infectious hematopoietic necrosis virus (IHNV) (LaPatra *et al.*, 1989) and red sea bream iridovirus (Nakajima *et al.*, 1995).

5.5 Enzyme-Linked Immunosorbent Assay (ELISA) is one of the most effective and popular immunoassays for measuring antigens, antibodies, proteins, etc. in various biological samples (measured in 96-well plates), making it possible to analyze several samples in a single experiment. There are four essential phases in the ELISA protocol: The following are the steps involved: The steps involved in this process are as follows: i) coating the ELISA plate with the capture antibody, which is an antibody raised against the antigen of interest; ii) adding samples (like urine, serum, or cell supernatant) so that any antigen of interest present in the sample will bind to the capture antibody; iii) adding secondary antibody that has been labeled with an enzyme (like horse radish peroxidase or alkaline phosphatase) that binds to the target antigen that has already bound to the plate through the variable region; and iv) adding a substrate, which reacts with the enzyme bound to the secondary antibody and causes the colour to develop (British Society for Immunology). The developed colour was directly proportional to the target antigen present, and the amount was measured using a plate reader or modified spectrophotometer. Based on the application of ELISA, can be divided into three major classes:

5.5.1 Antibody capture assays: An unlabelled antigen is immobilized on a solid phase as part of the standard process, and the antibody is allowed to attach to the antigen. A labelled secondary reagent (goat anti-mouse or goat anti-rabbit IgG antibody), which contains a conjugated enzyme that can precisely recognize the main antibody, can be used to directly label or identify the antibody. Finally, an enzyme-specific substrate was introduced to induce colour development. An ELISA plate reader or modified spectrophotometer was used to detect the intensity of the colour created (de la Peña and Leobert D 2001).

5.5.2 Antigen capture assays The primary functions of the assay are antigen detection and quantification. An unlabeled antibody was immobilized on the solid phase as part of the standard technique. After mixing a sample of the labelled antigen with a test solution containing an unknown amount of antigen, the combination was applied to the linked antibody, where the antigen of the test solution competed with the tagged antigen. The specified substrate was introduced after washing to measure the enzymatic activity of the bound material. Following the development, colour was assessed using either a customized spectrophotometer or an ELISA plate reader (de la Peña and Leobert D 2001).

5.5.3 Two-antibody sandwich assays: The main purpose of these tests was to determine the antigen content of the unidentified samples. Two antibodies that attach to non-overlapping epitopes on the antigen were required for the experiment. A purified antibody was immobilized on a solid phase using a standard approach, and a sample solution containing an unidentified antigen was allowed to bind. A secondary labelled antibody that has been chemically coupled to the enzyme horseradish peroxidase was allowed to adhere to the antigen after washing (to eliminate unbound antigens). Ultimately, an ELISA plate reader or a customized spectrophotometer is used to detect the intensity of the colour formed after adding a particular substrate against the enzyme (de la Peña and Leobert D 2001).

Conclusion

The aquaculture sector has experienced a revolution thanks to the use of cutting-edge methods for diagnosing fish diseases. These tools have made pathogen identification more precise, effective, and quick. Our capacity to identify and analyze a broad spectrum of fish pathogens has improved due to advancements in imaging technologies, bioinformatics, and advanced molecular tools like PCR, RFLP, AFLP and DNA microarrays. These instruments not only help in the early detection and treatment of fish diseases, which lowers their financial impact, but they also aid in the research and development of focused medications and vaccines. Furthermore, combining these technologies with artificial intelligence and data analytics may be able to forecast disease outbreaks and enhance fish health management in general.

References

Altinok, I., Kurt, I., 2003. Molecular diagnosis of fish diseases: a review. Turk. J. Fish. Aquat. Sci. 3, 131-138.

Bondad-Reantaso, M.G., Subasinghe, R.P., Arthur, J.R., Ogawa, K., Chinabut, S., Adlard, R., et al., 2005. Disease and health management in Asian aquaculture. Vet. Parasitol. 132, pp.249-272.

British Society for Immunology: https://www.immunology.org/public-information/bitesized-immunology/experimental-techniques/enzyme-linked-immunosorbent-assay (accessed on 28.12.2023)

Bullock GL, Griffin BR, Stuckey HM. 1980. Detection of Corynebacterium salmonius by direct fluorescent antibody test. Canadian Journal of Fisheries and Aquatic Sciences 37, pp.719-721

Callahan, D., 1973. The WHO definition of'health'. Hastings Center Studies, pp.77-87.

Cui, C., Shu, W. and Li, P., 2016. Fluorescence in situ hybridization: cell-based genetic diagnostic and research applications. Frontiers in cell and developmental biology, 4, p.89.

de la Peña, L.D., 2001. Immunological and molecular biology techniques in disease diagnosis. In Health management in Aquaculture (pp. 137-158). Aquaculture Department, Southeast Asian Fisheries Development Center.

de la Peña, L.D., 2002. Polymerase chain reaction (PCR) in disease diagnosis.

de la Peña LD, Momoyama K, Nakai T, Muroga K. 1992. Detection of the causative bacterium of vibriosis in kuruma prawn, Penaeus japonicus. Fish Pathology 27, pp.223-228

González, S.F., Krug, M.J., Nielsen, M.E., Santos, Y. and Call, D.R., 2004. Simultaneous detection of marine fish pathogens by using multiplex PCR and a DNA microarray. Journal of clinical microbiology, 42(4), pp.1414-1419.

Hamera, B., and Bondad-Reantaso, M.G., 2001. Food and Agriculture Organization of the United Nations and Network of Aquaculture Centres in Asia-Pacific, 2001. In: Bondad-Reantaso, M.G., McGladdery, S.E., East, I., Subasinghe, R.P. (Eds.), Asia Diagnostic Guide to Aquatic animal Diseases.FAO and NACA, Rome.

Huang, T.S., Cerenius, L. and Soderhall, K. 1994. Analysis of genetic diversity in the crayfish plagues fungus, Aphanomyces astaci, by random amplification of polymorphic DNA. Aquaculture, 126, pp.1–10.

Ke, R, Mignardi, M, Hauling, T and Nilsson, M., 2016. Fourth Generation of Next-Generation Sequencing Technologies: Promise and Consequences. Human mutation 37, pp.1363-1367

LaPatra SE, Roberti KA, Rohovec JS, Fryer JL. 1989. Fluorescent antibody test for rapid diagnosis of infectious hematopoietic necrosis. Journal of Aquatic Animal Health 1. Pp.29-36

Lewis DH, Savage NL. 1972. Detection of antibodies to Aeromonas liquifaciens in fish by an indirect fluorescent antibody technique. Journal of the Fisheries Research Board of Canada 27, pp.1389-1393

Lightner, D.V., 1996. A Handbook of Pathology and Diagnostic Procedures for Diseases of Shrimps. Special Publication. World Aquaculture Society, Baton Rouge, LA.

Lightner, D.V., Redman, R.M., 1998. Shrimp diseases and current diagnostic methods. Aquaculture 164, pp.201-220.

Lilley, J.H., Cerenius, L. and Soderhall, K. 1997. RAPD evidence for the origin of crayfish plague outbreaks in Britain. Aquaculture, 157, pp.181-185.

Maxam, A. M, & Gilbert, W., 1977. Proc. Natl. Acad. Sci. USA 74, pp.560-564

Microbiology: Canadian edition: https://ecampusontario.pressbooks.pub/microbio/chapter/fluorescent-antibody-techniques/ (accessed on 28.12.2023)

Miller, B. M., metz, D., schmid, J., rudin, P. M., & blumenthal, M. S. (2021). Measuring the Value of Invention Published by the RAND Corporation, Santa Monica, Calif.

Mishra, S.S., Das, R., Sahoo, S.N. and Swain, P., 2020. Biotechnological tools in diagnosis and control of emerging fish and shellfish diseases. In Genomics and biotechnological advances in veterinary, poultry, and fisheries (pp. 311-360). Academic Press.

Nakajima K, Maeno Y, Fukudome M, Fukuda Y, Tanaka S, Matsuoka S, Sorimachi M., 1995. Immunofluorescence test for the rapid diagnosis of red sea bream iridovirus infection using monoclonal antibody. Fish Pathology 30, pp.115-119

OIE, 2000. Diagnostic Manual for Aquatic Animal Diseases, third ed. Office International des Epizootics, Paris.

Pingoud, A. and Jeltsch, A., 2001. Structure and function of type II restriction endonucleases. Nucleic acids research, 29(18), pp.3705-3727.

Sanger, F. & Coulson, A. R., 1975. J. Mol. Blol. 94, pp.441-448

Sanger, F., Nicklen, S. and Coulson, A.R., 1977. DNA sequencing with chain-terminating inhibitors. Proceedings of the national academy of sciences, 74(12), pp.5463-5467.

Scully, J.L., 2004. What is a disease? Disease, disability and their definitions. EMBO reports, 5(7), pp.650-653.

Shalon, D.D., Smith, S.J. and Brown, P.O. 1996. A DNA microarray system for analyzing complex DNA samples using two-color fluorescent probe hybridization. Genome Res., 6, pp.639–645.

Subasinghe, R.P., Bondad-Reantaso, M.G., McGladdery, S.E., 2001. Aquaculture development, health and wealth. In: Subasinghe, R.P., Bueno, P., Phillips, M.J., Hough, C., McGladdery, S.E., Arthur, J.R. (Eds.), Aquaculture in the Third Millennium. Technical Proceedings of the Conference on Aquaculture in the Third Millennium, Bangkok, Thailand, 20-25 February 2000. NACA/FAO, Bangkok/Rome, pp. 167-191.

Toranzo AE, Baya AM, Roberson BS, Barja JL, Grimes DJ, Hetrick FM. 1987. Specificity of slide agglutination test for detecting bacterial fish pathogens. Aquaculture 61, pp.81-97

Vaneechoutte, M., 1996. DNA fingerprinting techniques for microorganisms: a proposal for classification and nomenclature. Molecular biotechnology, 6, pp.115-142.

Şahin, E.Ç., Aydın, Y., Gilles, T., Uncuoğlu, A.A. and Lucas, S.J., 2022. Concepts and applications of bioinformatics for sustainable agriculture. In Bioinformatics in Agriculture (pp. 455-489). Academic Press.

8

Fish Therapeutics and Its Limitations

Samikshya Mishra** and Debiprasad Kantal

ICAR-Central Institute of Fisheries Education, Mumbai, Maharashtra (400061), India

Abstract

Aquaculture is a substantial economic enterprise and thriving sector with diverse resources and potential. They play a critical role in supplying food, nutrition, and employment opportunities. Fish therapeutics play a crucial and indispensable role in the overall function in maintaining the health of fish populations, contributing to the stability and productivity of the industry, while minimizing the impact of diseases on both farmed and wild fish populations. This article explores the diverse techniques employed in fish health management, delving into their applications and advantages while simultaneously addressing the inherent limitations that pose challenges to sustainable aquaculture practices. In addition, this article discusses the regulatory landscape surrounding fish therapeutics and the challenges faced by aquaculturists in adhering to diverse and often ambiguous regulations.

Keywords: *Therapeutics, Drugs, Medicine, Aquaculture, Limitations.*

1. Introduction

Aquaculture is a thriving industry that contributes significantly to the economy by creating employment, sustenance, and nutrition owing to diverse resources and possibilities. Aquaculture is experiencing rapid growth as the swiftest food production sector is expanding, with an annual increase exceeding 8%. Fish production reached an estimated level of 16.25 million metric tons (MMT) (MPEDA). In the dynamic realm of aquaculture, the application of fish therapeutics is important for maintaining the health and productivity of fish populations. As the industry strives to meet the escalating global demand for seafood, understanding the nuances of fish therapeutics is crucial. This article explores the diverse techniques employed in fish health management, delving into their applications and advantages while simultaneously addressing the inherent limitations that pose challenges to sustainable aquaculture practices. Navigating through these complexities is essential to forge a path towards effective, ethical, and environmentally responsible fish therapeutics in the ever-evolving landscape of aquaculture.

Aquaculture development is currently focused on increasing vertical expansion by intensifying cultural practices, with a limited scope for horizontal expansion. The aquaculture industry has expanded vertically as a result of the introduction of various species and higher stocking densities, which has increased the incidence of disease outbreaks, frequently leading to widespread mortality and decreased overall production. Aquaculture has been negatively affected in recent years because of frequent disease outbreaks, primarily as a result of boosted culture systems for greater economic advantage (Walker and Winton *et al.*, 2010). Greater diversity and larger quantity of species have also led to the development of new cultural systems and practices, potentially affecting the emergence and spread of infections. (Murray *et al.*, 2005). Fish therapeutics have become essential for overcoming these challenges in the aquaculture industry (Nayak *et al.*, 2007; Sahoo *et al.*, 2013). Fish diseases and the use of aqua medicines in India and other Asian countries (Bondad-Reantaso *et al.*, 2005; Faruk *et al.*, 2004) have increased the control of production losses (Burridge *et al.*, 2010; Ali *et al.*, 2014; Chowdhury *et al.*, 2012). Therapeutants in aquaculture are used for pond preparation, soil and water management, feed additives, growth, and augmentation of natural aquatic productivity (Alam *et al.*, 2015). Drugs and chemicals used in India's grow-out farms and hatcheries can be broadly classified as follows: The current understanding is insufficient for the development of comprehensive disease management strategies. Many infectious diseases (bacterial, fungal, and parasitic) in fish lack effective vaccinations; thus, safe and effective management strategies to minimize losses due to outbreaks are urgently needed. There are very few FDA-approved therapeutic agents available, the sector is constantly confronted with disease outbreaks, and there are no effective strategies to reduce losses.

2. Importance of Fish Therapeutics in aquaculture and fisheries management

Fish therapeutics are crucial in aquaculture and fisheries management, as they safeguard the health of fish populations. Effective disease management ensures sustainable production, prevents economic losses for aquaculturists, and maintains an ecological balance. By employing various therapeutic techniques, such as vaccinations, antibiotics, and chemical treatments, aquaculture practitioners can enhance the overall well-being of fish stocks, contributing to the stability and productivity of the industry while minimizing the impact of diseases on both cultivated farm and wild fish populations.

Ensuring the health of fish populations and the long-term viability of the aquaculture and fisheries sectors requires a holistic approach to fish health management. Good husbandry practices encompass various aspects, including maintaining high water quality, effective system management, and providing proper nutrition for fish. These practices not only facilitate good health, but also play a significant role in preventing disease outbreaks. Biosecurity is a fundamental aspect of fish health management. It is essential to prevent the introduction and spread of infectious diseases in aquaculture facilities and in the broader environment. Effective biosecurity measures are crucial for minimizing the risk of both the introduction of diseases into a facility and the potential spread of infected animals or agents to other sites. This includes practices such as good animal management, sourcing and maintaining healthy stocks, robust pathogen management, and educating staff and visitors to ensure their contributions to disease prevention and management. By implementing these principles of good husbandry and biosecurity, industry can significantly reduce the occurrence and impact of disease outbreaks. Moreover, this proactive approach to fish health management contributes to the sustainability and long-term viability of aquaculture and fisheries. Aquaculture uses a variety of antimicrobial treatments, such as antibiotics, insecticides, anesthetics, hormones, pigments, minerals, and vitamins to prevent or manage infections (Figure1).

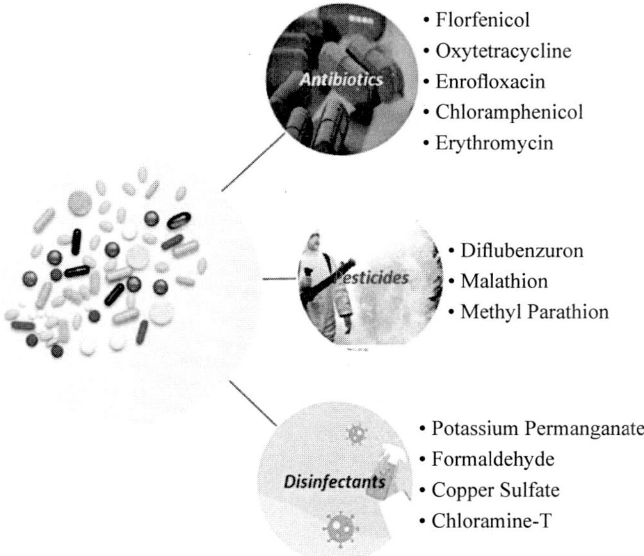

Fig. 1. Commonly used fish antioxidants in aquaculture.

3. Aquaculture Therapeutics Market Outlook

The market for aquaculture therapy is anticipated to expand at a compound annual growth rate (CAGR) of 8.6% from 2023 to 2033. The market is expected to be worth $1.74 billion in 2023, having a share of $3.98 billion 2033. The need for an effective aquaculture therapy is expected to increase the demand for aquaculture therapies. Furthermore, the increasing occurrence of bacterial diseases in aquaculture species supports market expansion. Growing global fish consumption and technological developments in aquaculture output are some of the factors that propel aquaculture therapeutic sales throughout the forecast period.

The list of medications available in the market includes hydrogen peroxide, tricaine methane sulfonate, florfenicol, oxytetracycline, acetic acid, formalin solution, sulfamethazine/ormetoprim combination and calcium chloride. Aquaculture treatments are expected to be in high demand throughout the projected period because these drugs treat a number of ailments that are damaging to both consumers and aquatic animals.

Despite the lengthy history of aquaculture practices in some countries, the aquaculture therapeutics market is currently expanding, owing to numerous growth opportunities. Furthermore, as humans consume fish or animals, the FDA has established specific regulatory criteria for medications delivered to fish in aquaculture, where they are classified as conditionally approved and indexed animal pharmaceuticals.

4. Methods of drugs administration in aquaculture

Antimicrobials are commonly administered in aquaculture through two main methods: medicated feed and water medicine. Fish are treated through a water bath when their biomass is low, as in the case of fry, making oral therapy ineffective. This is a simple process that requires only the volume of water to calculate the final concentration of the drug. Because low-molecular-weight drugs can be evenly distributed throughout water, they are typically recommended for use as water pharmaceuticals. Medication is absorbed by fish and acts on pathogens via the skin, mucosa, gills, and epithelia. Medicated feeds are made by combining a small amount of antimicrobial medication with an extruded and homogenized diet; alternatively, antibiotics can be sprayed or coated on top of the feed. In contrast to water medicine, the advantages of in-feed medication include less wastage. It also reduces the unfavorable exposure of the drug to the environment and other fish. Treating a large number of unhealthy fish in this manner is standard practice. The only constraint of this procedure is that the treated fish must be actively fed. As the specific amount is known, the administration approach is employed in experimental work instead of oral delivery. In experimental studies, the gavage method is employed for oral administration because the exact dosage is known. During this treatment, a syringe filled with medication was injected into the stomach of the fish through a stomach tube. Prior to administering the medication, the fish were anesthetized. This method is labor-intensive and stressful for fish; therefore, it is rarely used in aquaculture. Injection techniques are frequently used to administer vaccines and treat a limited number of fish or significant species, although they are time-consuming. Common Injection methods for fish include intramuscular, intraperitoneal, and dorsomedian sinuses.

5. Commonly applied Therapeutics in aquaculture

Defoirdt *et al.* (2011) reported that aquaculture has experienced an increase in the occurrence of bacterial infections due to intensification, which has resulted in greater use of antimicrobial drugs. Antibiotics are frequently administered to groups of ill fish for short periods of time. Aquaculture has been permitted for the use of florfenicol, sulfamerazine, oxytetracycline hydrochloride, oxytetracycline dihydrate, and a combination of sulfadimethoxine and ormetoprim, provided that the fish have less residuals than a specified daily limit. Other vaccines that are readily accessible for use in aquaculture protect against a range of bacterial and viral diseases. According to Woo (2006), single-celled protozoans, multicellular trematodes, crustaceans, and arthropods can cause parasitic infestations in freshwater fish. To address parasitic infestations, aquaculture regularly utilizes drugs and chemicals such as Nuvan, Emamectin Benzoate (EB), Cliner, Butox Vet, Ectodel (2.8%), Paracure-IV, Hitek Powder, and others. The market demand for Butox Vet and Cliner is higher than that for the others. Although several of these items have licences for use in veterinary animals, there are no specific guidelines for their use in aquaculture. An inventory of several USFDA-approved medications used in aquaculture is presented in Table 1.

5.1 Products for the treatment of Water and Soil

Pond preparation is the first step towards increasing fish productivity. Pond water should be treated before stocking fish to mineralize organic materials, correct the pH, and disinfect. In this case, lime in the form of limestone (CaCO3), slaked lime (Ca (OH)2), or unslaked lime (CaO) is utilized. Aqua medicines, such as zeolite, bio-aqua, geotox, lime, aquanone, zeocare, and zeo prime, are used for pond preparation and water quality control. Dried ponds can also be sterilized and disinfected with potassium permanganate (KMnO4) or active iodine. To revitalize the soil and regulate algal development while absorbing fouling elements, zeolite, which is commercially available and aluminum silicate are combined with lime.

5.2 Disinfectants Employed in Aquaculture

Disinfectants are widely used in several aquaculture sectors worldwide. The majority of these are used in intensive culture, specifically in grow-out culture systems. They work on keeping things hygienic, setting up the manufacturing cycle and machinery, and sometimes even treating infections. Bleaching, BKC, EDTA, aquakleen, efinol, and formalin are examples of medications used for disease treatment. The protozoan infections were also treated with formalin. Benzalkonium Chloride (BKC) is the most frequently used medication to treat bacterial illnesses, and efinol can be used to increase stress tolerance. Common disinfectants are sodium hypochlorite, BKC, calcium carbide, and Na-EDTA. They are most frequently used in hatcheries, whereas they are occasionally used in grow-out ponds.

The first step is to thoroughly clean the tanks and equipment, ensuring that all dirt and organic material are removed. Pathogens can be protected from disinfection by organic materials such as biofilms and soil. Liquid household bleach (3–6% NaHClO) has a concentration of 200 mg/L when used at 35 mL per gallon of water. Commercial farms commonly use granular bleach (HTH; calcium hypochlorite), which is more stable than household bleach. There are three available concentrations of accessible chlorine: 15, 50, and 65%. For disinfection, 200 mg/L chlorine should be used for at least an hour.

Mycobacteria have a waxy cell wall, which makes them resistant to bleach disinfection. Mycobacteria should be successfully eliminated by applying alcohol to contact surfaces and equipment after bleach treatment. With active components of 21.4% potassium peroxymonosulfate and 1.5% sodium chloride, another popular disinfectant for aquaculture is extremely efficient at concentrations of 1–2% and significantly less harmful to fish (manufacturer guidelines must be followed). Benzalkonium chloride and other quaternary ammonium compounds work well as disinfectants and can be used at concentrations of 500 mg/L concentrations for an hour. Subsequently, the remaining films were thoroughly removed by washing. Virkon Aquatic and quaternary ammonium compounds are more effective for disinfection and net dips. Peracetic acid-containing compounds have recently become available for use as disinfectants in aquaculture.

5.3 Piscicides and herbicides

Piscicides and herbicides are used for specific purposes in aquaculture. must be used cautiously because of their potential impacts on ecosystems and non-target aquatic plants. They should be applied in a manner that minimizes any adverse effects, following label guidelines and regulations. Piscicides are chemicals used to eliminate unwanted fish species from water bodies. Herbicides are used to control aquatic plants that can compete with cultured organisms for nutrients and light, interfere with water flow, oxygen levels, and other factors important for the health of aquaculture systems, and harbor pests and pathogens that could affect fish populations. The safest and most approved herbicide for use in fishponds is probably flouride. Its aftereffects might not become apparent for 30–90 days, and its residue might last for two–12 months. Some glyphosate formulations contain chemicals that are hazardous to aquatic life; glyphosate is used to suppress emergent and shoreline weeds. Although 2,4-D has been successful in managing aquatic vegetation, its toxicity increases as the pH decreases. At pH levels above 8, they are less efficient, whereas in acidic water ($pH < 6$), they are more hazardous. Diquat is a long-standing treatment for fish columnar disease that was first reported as a contact herbicide (pesticide).

5.3.1 Potassium Permanganate

Potassium permanganate is used in aquaculture as a disinfectant and a parasiticide. It is effective in treating a variety of external infections in fish, including parasites, bacteria, and fungi. In order to control algae and lower the biological oxygen demand, it is also utilised in the management of water quality. The dose of potassium permanganate can vary widely based on the condition being treated, water temperature, organic matter content, and water chemistry, such as hardness and pH. Owing to its potent oxidizing properties, it must be used with caution, and accurate dosing is crucial to avoid toxicity to fish. Potassium permanganate is usually applied as a bath treatment, and fish are exposed to a diluted solution for a set period. It is a potent oxidizing chemical that "burns" organic material from the exterior of fish. Fish die from overuse, especially from repeated applications in a short amount of time (more than once per week). Typically, a KMnO4 concentration of 1-2 mg/L is safe and efficient. The permanganate requirement is higher in water with high organic content. If the required concentration of KMnO4 exceeded 6 mg/L, the organic load was too high. Greater salinity in the water causes it to be more harmful.

5.3.2 Copper Sulfate

For many years, CuSO4 has been utilized as a parasiticide; owing to its low cost, it is particularly helpful in large production ponds. The safe use of copper necessitates knowledge of the total alkalinity, the volume of water to be treated, and any potential species sensitivity, because copper is extremely harmful to fish. Copper sulfate is often used in aquaculture as a disinfectant and a parasiticide. It is effective in treating various parasitic and fungal infections, and is sometimes used to control algal growth within the aquaculture environment. However, it must be used with caution because of its toxicity in fish at higher concentrations and its potential environmental impact. Appropriate dosing is crucial to minimize harmful effects. The application should be dependent on the total alkalinity (TA) of the water, as CuSO4 concentrations in freshwater systems are thought to be 100% active. The use of copper sulfate is not safe if TA is less than 50 mg/L. When TA ranges from 50 mg/L to 250 mg/L, a rapid mortality algal bloom can lead to a disastrous oxygen shortage. If a pond is heavily algal-blighted, the use of CuSO4 could be dangerous. To minimize harmful consequences, this enables the physiology of the fish to adjust and initiate detoxification processes. Cu is extremely hazardous to many invertebrates, and even after slow adaptation, some marine species remain vulnerable.

5.3.3 Diflubenzuron

Crustacean parasites are controlled by the insecticide diflubenzuron, which has a restricted usage label and is a chitinase inhibitor. Dimilin is commercially approved for the management of anchorworms on ornamental fish in ponds, tanks, and bait fish. Diflubenzuron should not be used to produce fish for human consumption.Anesthetics/Sedatives

5.3.4 Tricaine Methanesulfonate

The only fish anesthetic approved by the FDA is tricaine methanesulfonate. It was derived from benzocaine. A permission has granted to TMS to temporarily immobilize fish and other cold-blooded aquatic species. TMS can only be used on Salmonidae, Ictaluridae, Percidae and Esocidae fish species intended for human consumption, and the water temperature must not exceed 10°C (50 °F). The withdrawal period was 21 d.

5.3.5 Eugenol

Eugenol, derived from clove oil, has gained recognition as a safe and effective fish anesthetic for use in both freshwater and marine fisheries with minimal risk of adverse effects. Its potential benefits in fish anesthesia contribute to the overall welfare and health management of fish in aquaculture and fisheries.

5.4 Antibiotics and Antimicrobial agents

These agents are used to treat bacterial infections, common antibiotics include florfenicol oxytetracycline, sulfonamides etc., Aquaflor is an approved medication feed containing florfenicol for use against specific pathogens in coldwater disease (Flavobacterium psychrophilum) in salmonids, enteric septicemia (Edwardsiella ictaluri) in channel catfish, furunculosis (Aeromonas salmonicida) Salmonids raised in freshwater, warmwater finfish raised in freshwater that have streptococcal septicemia, and columnaris diseases. It has also been promoted as a VFD medicine. When used against gram-positive streptococci and gram-negative bacteria, this broad-spectrum antibiotic works well.

Oxytetracycline and enrofloxacin are used to treat a broad range of bacterial infections; florfenicol is used to treat respiratory and systemic bacterial infections; sulfonamides are used to inhibit bacterial growth; and trimethoprim-sulfamethoxazole, which is a combination antibiotic, treats bacterial infections in fish. One approved in-feed VFD medication is oxytetracycline dihydrate: 1. for the control of mortality in freshwater-reared salmonids induced by F. psychrophilum coldwater disease, 2. to reduce mortality in freshwater-reared rainbow trout caused by F. columnaris disease, and 3. to mark skeletal tissue in Pacific salmon.

Ormetoprim sulfadimethoxine is an FDA-approved in-feed therapy for salmonid furunculosis (salmonicida). The FDA has approved the use of 35% aquaculture-grade hydrogen peroxide in fish to combat bacterial gill disease in salmonids (caused by F branchiophilum), external columnaris in freshwater-reared cold-water finfish and catfish(caused by F columnare), and fungal infection of finfish eggs (saprolegniasis) raised in freshwater, which are typical external infections in fish that can occur as a result of handling, excessive organic load, or other water quality disturbances in the affected species.

Hydrogen peroxide is utilized as a short-term, continuous-flow bath. Three treatments are administered daily or on consecutive alternate days, which are used for treating both fungal and bacterial infections, povidine-iodide applied topically for disinfecting wounds, and formalin effective against external parasites, bacteria, and fungi. Quaternary ammonium compounds used for disinfecting equipment, tanks and water treatments. Salt is used to control parasites and regulate osmoregulation in fish. Copper-based compounds have been used to control external parasites and fungal infections.

5.5 Vaccines

Administration of vaccines to protect against specific diseases. The principles behind vaccination are memory and specificity. The ability of the immune system to remember previously encountered pathogens produces a stronger immune response. Vaccination boosts immunity and decreases the severity of infection. Vaccines are biological preparations that stimulate the immune system to identify and fight specific pathogens such as bacteria or viruses. They contain a weekened or inactivated form of the entire targeted organism or parts thereof. Some types of vaccines are mentioned below-

- Live attenuated vaccines include a suspension of attended live pathogens that can replicate and produce protective immune responses inside the host body, but are unable to cause disease. Edwardsiella ictaluri attenuated live vaccines have been licensed in the USA.

- Inactivated vaccines contain heat-killed or chemically killed pathogens. Bacterins are vaccines obtained from inactivated bacteria and are commonly used in salmonids against bacteria such as Aeromonas salmonicida.
- Subunit vaccines contain specific parts of a pathogen, mostly proteins or sugars. These can be prepared in the laboratory using genetic engineering techniques. A subunit vaccine used in Chile against infectious pancreatic necrosis virus in salmon is Intervert's Compact® IPN.
- Recombinant vaccines: In these vaccines, harmless parts of the genetic material from disease-causing microorganisms are inserted into the carrier or vector, which are weakened bacteria. Application in the prevention of diseases such as viral hemorrhagic septicemia.
- DNA vaccines: Developed by introducing genetic material directly into fish cells to produce targeted proteins. Used against various fish viruses.

5.6 Utilization of Probiotics in Aquaculture

Live bacterial cultures and extracellular fermentation products are the most frequently used probiotics for pond management. Probiotics are beneficial for the management of aquaculture ponds. They help improve organic matter decomposition, reduce nitrogen and phosphorus concentrations, promote algal growth, increase dissolved oxygen availability, control ammonia, and lower disease incidence, while enhancing fish survival and production.

However, studies have indicated that the addition of probiotics has relatively few favorable effects (Boyd and Gross, 1998; Queiroz *et al.*, 1998). Probiotics primarily comprise various concentrations of beneficial bacteria such as Bacillus sp., Streptococcus faecalis, Rodobacter sp., and Rodococcus sp. The addition of probiotics to aquaculture ponds is not expected to pose any food or environmental safety risk.

5.7 Feed additives

Advances in aquaculture techniques have led to a considerable shift away from the use of agricultural byproducts as supplementary fish diets and toward nutritionally adequate species-specific complete feeds. vitamins, colors, chemo-attractants and preservatives such as mold inhibitors and antioxidants are all included in these balanced meals. There is a specific reason why these feed additives are included in diet formulas, despite their often-varying compositions and qualities. Certain additives, such as exogenous enzymes and acidifiers, are utilized to increase feed digestibility or reduce the negative effects of anti-nutrients in order to enhance fish performance. Other substances, such as phytogenics, probiotics, prebiotics, and immune stimulants, have been designed to improve gut health, resistance to stress, and immunity to diseases.

5.8 Growth promoters for fish

There are various chemicals found in chemical stores that are utilised as growth enhancers, such as AQ Grow-L, Megavit Aqua, AQGrow-G, F Aqua, Fish vita plus, Aqua Boost, Aqua Savour, ACmix, Nature Aqua GP Fibosoel, Vitamin premix, Grow fast, Orgavit auqa, Vitamix, and many more. Aqua Boost contains immune stimulants that boost non-specific immunity in fish.

Table 1. Approval Status and Withdrawal Periods of Aquaculture Drugs

Drug	Trade name	Generic Use	Mode of Application	Withdrawal time
Copper sulfate (CuSO4)	-	Parasiticide (gills and skin)	Immersion	7 days
Potassium permanganate (KMnO4)	-	Parasiticide (gills and skin)	Immersion	7 days
Tricaine methanesulfonate	Tricaine-S®	Anesthetic	Immersion	21 days
Eugenol	Aqui-S20E®	Anesthetic	Immersion	72 hours
Diflubenzuron	Dimilin®	Parasiticide (external crustaceans)	Immersion	-
Florfenicol	Aquaflor®	Antimicrobial	Medicated feed	15 days
Oxytetracycline dihydrate	Terramycin-200®	Antimicrobial	Medicated feed	21 days
Hydrogen peroxide (35%)	PEROXAID®	Antimicrobial (gills and skin)	Immersion	-
Chloramine T	HALAMID Aqua®	Antimicrobial (gills and skin)	Immersion	
Diquat	Reward®	Herbicide Antimicrobial (gills and skin)	Immersion	5 days
Formalin	Formalin-FTM Parasite-S® Formalin-B	Parasiticide (gills and skin)	Immersion	-
Chorionic gonadotropin	Chorulon®	Spawning aid	Injection	-
sGnRHa + domperidone	Ovaprim®	Spawning aid	Injection	-

6. Limitations of Fish Therapeutics

6.1 Environmental Impact

The chemicals and pharmaceuticals used in fish therapeutics can adversely affect aquatic ecosystems. Residues from treatments may persist in water bodies, leading to unintended consequences for nontarget organisms. This section explores the environmental impact of fish therapeutics, emphasizing the need for sustainable practices to mitigate ecological damage.

6.2 Development of Resistance

The emergence of resistant strains of pathogens and antibiotic-resistant fish may have serious effects on the long-term effectiveness of fish therapeutics. An in-depth analysis of the factors contributing to resistance, such as the improper use of antibiotics and inadequate treatment protocols, sheds light on the urgency of adopting alternative strategies to manage fish diseases.

6.3 Regulatory Challenges

The lack of standardized guidelines and international cooperation in regulating fish therapeutics present significant challenges. This section investigates the regulatory landscape surrounding fish therapeutics and discusses the obstacles aquaculturists face in adhering to diverse and often ambiguous regulations.

6.4 Cost and Accessibility

The economic viability of fish therapeutics is a critical factor for aquaculturists, particularly in developing countries. The high costs associated with pharmaceuticals and treatments may limit accessibility, leading to potential health crises in fish populations. This section explores the economic challenges and discusses alternative cost-effective solutions for sustainable fish health management.

Conclusion

Ensuring fish health in aquaculture and fishery management is of paramount importance. However, the limitations and challenges associated with current practices highlight the need for a paradigm shift towards more sustainable and effective solutions. As the aquaculture industry continues to evolve, research and development efforts must address the environmental, regulatory, and economic aspects of fish therapeutics to ensure a healthy and resilient aquatic ecosystem for future generations.

References

Alam CA, Uddin MS, Vaumi S, Abdulla AA, Aquazdrugs, chemicals used in aquaculture of Zakigonj upazilla, Sylhet. Asian Journal of Medical and Biological Research. 2015; 1:336-349.

Ali MM, Rahman MA, Hossain MB, Rahman MZ. Aquaculture Drugs Used for Fish and Shellfish Health Management in the South western Bangladesh. Asian Journal of Biological Sciences. 2014; 7:225-232.

Bondad-Reantaso MG, Subasinghe RP, Arthur JR, Ogawa K, Chinabut S, et al Disease and health management in Asian aquaculture. Vet Parasitol. 2005; 132:249-272.

Boyd CE, Gross A. Use of probiotics for improving soil and water quality in aquaculture ponds. In: Flegel, T.W. (Ed.), Advances in Shrimp Biotechnology. Proceedings to the Special Session on Shrimp Biotechnology, 5th Asian Fisheries Forum, 11-14 November 1998, Chiengmai, Thailand. The National Center for Genetic Engineering and Biotechnology, Bangkok, Thailand, 1998, 101-106

Burridge L, Weis JS, Cabello F, Pizarro J, Bostick K. Chemical use in salmon aquaculture: A review of current practices and possible environmental effects. Aquaculture. 2010; 306:7-23

Chowdhury AKJ, Saha D, Hossain MB, Shamsuddin M, Minar MH. Chemicals Used in Freshwater Aquaculture with Special Emphasis to Fish Health Management of Noakhali, Bangladesh. African Journal of Basic & Applied Sciences. 2012; 4:110-114.

Costello BMJ, Grant A, Davies IM, Cecchini S, Papoutsoglou S, *et al.* The control of chemicals used in aquaculture. European Journal of Applied Ichthyology. 2001; 17:173-180

Defoirdt T, et al, Alternatives to antibiotics to control bacterial infections: luminescent vibriosis in aquaculture as an example, In: Trends in biotechnology. 2007; 25(10):472-9

Defoirdt T, Sorgeloos P, Bossier P. Alternatives to antibiotics for the control of bacterial disease in aquaculture, In: Current opinion in microbiology. 2011; 14(3):251-8.

Faruk MAR, Alam MJ, Sarker MMR, Kabir MB. Status of fish disease and health management practices in rural freshwater aquaculture of Bangladesh. Pakistan Journal of Biological Science. 2004; 7:2092-2098.

Murray AG, Peeler EJ. A framework for understanding the potential for emerging diseases in aquaculture. Prev Vet Med. 2005; 67:223-235

Nayak SK, Swain P, Mukherjee SC. Effect of dietary supplementation of probiotic and vitamin C on the immune response of Indian major carp, Labeo rohita (Ham.). Fish Shellfish Immunol. 2007; 23:892-896

Queiroz JF, Boyd CE, Gross A. Evaluation of a bioorganic catalyst in channel catfish, Ictalurus punctatus, ponds. J Appl. Aquac. 1998; 8:49-61.

Queiroz JF, Boyd CE. Effects of a bacterial inoculum in channel catfish ponds. J World Aquac. Soc. 1998; 29:67- 73.

Sahoo PK, Mohanty J, Garnayak JSK, Mohanty BR, Kar B. *et al.* Estimation of loss due to argulosis in carp culture ponds in India. Indian Journal of Fisheries, 2013

Walker PJ Winton JR. Emerging viral diseases of fish and shrimp. Vet Res. 2010; 41:51.

Woo PTK. Fish Diseases and Disorders, (2nd edn). Protozoan and Metazoan Infections. Cambridge, USA, 2006.

9

Application of Herbs in Fish Feed and its Importance and Challenges

Debiprasad Kantal **and** ***Samikshya Mishra***

ICAR-Central Institute of Fisheries Education, Mumbai-400061, Maharashtra, India

Abstract

The aquaculture sector in India faces challenges such as low production due to expensive feeds, disease outbreaks, and poor broodstock development. Small-scale farmers, lacking resources, struggle with the high costs of commercial diets and chemical additives essential for steady production. Moreover, safety concerns and prohibitions on chemical additives in aquaculture have prompted a need for alternatives. The incorporation of herbs and their extracts as feed additives emerges as a promising solution to enhance fish growth and disease resistance. Plant-derived secondary metabolites, known for their growth-promoting and immune-enhancing properties, offer an inexpensive, safe, and effective alternative to conventional chemotherapeutic additives. This chapter explores the application of herbs and their extracted metabolites in aquaculture, focusing on their roles as appetite stimulators, growth promoters, antimicrobials, antiparasitics, antioxidants, and immunostimulants in fish along with their mode of action, advantages, and limitations.

Keywords: *Herbs, Secondary Metabolite, Extraction, Phytobiotics.*

1. Introduction

Since the 1980s, the aquaculture industry has boomed due to growing market needs and limited natural harvesting options. To meet demand, high-density fish farming methods have become popular, but overcrowding can stress fish, making them prone to diseases. The implications of diseases in aquaculture extend beyond mere mortality and morbidity, encompassing poor growth rates, inferior flesh quality, and diminished profit margins. Consequently, a critical imperative arises to implement preventive and therapeutic measures.

Traditionally, chemotherapeutic agents, including antibiotics and disinfectants, have been employed for the treatment and prevention of various diseases in farmed fish. Nonetheless, their indiscriminate and continuous use raises concern, as it may foster the development of antibiotic-resistant bacteria, contribute to environmental pollution, and result in the presence of chemical residues in fish products. The existence of antimicrobial residues in food has garnered attention from the scientific community, given the associated risks to food safety and public health. These risks encompass not only antibiotic resistance (AMR) but also potential issues related to severe allergic reactions, carcinogenicity, renal dysfunction, mutagenicity, developmental toxicity, bone marrow depression, and disruption of the normal gut microbiota. Consequently, a reevaluation of preventive and therapeutic strategies in aquaculture is imperative to ensure sustainable practices without compromising human health (Bondad, Reantaso *et al.*, 2023). The unregulated utilization of antimicrobial agents in aquaculture poses potential risks to human health. Consequently, several countries have implemented prohibitions on specific chemotherapeutics and are reluctant to import aquaculture products treated with antibiotics and chemicals. In response to these concerns, researchers have intensified their endeavors to explore natural alternatives for developing dietary supplements that can promote the growth performance, health, and immune system of cultured fish. In this context, the utilization of herbs emerges as a promising solution (Syahidah *et al.*, 2015).

An herb is a type of plant that contains different natural compounds like flavonoids, pigments, and essential oils. These compounds give herbs their scent, taste, and potential health benefits. In recent years, researchers have been studying herbal medicines as possible additions to fish food to prevent and treat diseases (Valladão *et al.*, 2015). The active ingredients in herbs can help boost the immune system, improve metabolism, and enhance growth in aquatic animals. This suggests that using plants in fish food could be a good alternative to antibiotics, chemicals, vaccines, and other synthetic substances. The incorporation of herbs into fish feed aligns with the growing demand for eco-friendly and health-promoting aquaculture practices. The diverse bioactive compounds in herbs confer several advantageous properties, positively influencing fish health and contributing to overall aquaculture sustainability. Recent studies have explored the extended application of herbs due to their natural, harmless, easy-to-prepare, and cost-effective nature, coupled with minimal side effects on both fish and the environment (Pu *et al.*, 2017). These herbs can be employed in various forms, including the whole plant, specific plant parts (such as leaves, roots, or seeds), or as extract compounds, either independently or in combination with other feed additives. Application methods encompass incorporating them into the feed or utilizing them through fish immersion in treated water (Chang, 2000). Despite the numerous advantages of using herbs in aquaculture, there is a noticeable absence of comprehensive documentation outlining indigenous herbs, their benefits, and their applications. Hence, this chapter is dedicated to exploring the potential application of herbs in aquacultureparticularly emphasizing on indigenous and commonly found plants that can promote growth, immunity, and reproductive performance in various fish species.

2. Concept of phytobiotics

Phytobiotics are plant products that can be incorporated into feed to improve growth, immunity, and productivity. Phytobiotics have been known for its antioxidant, antimicrobial, analgesic, antiparasitic, immune booster, and growth promoter properties. The application of phytobiotics in aquaculture for improved growth, and non-specific immunity has already been studied (Asimi & Sahu, 2013).

2.1 Classification

2.1.1 Medicinal plants are generally classified on the basis of their growth habit (Asimi & Sahu, 2013).

2.1.1.1 Trees: Firm single stem supporting the entire structure.

2.1.1.2 Shrubs: Multiple stems forming a bush-like structure suitable for barricades.

2.1.1.3 Herbs: Lack of a firm stem, with a flexible and juicy structure, devoid of woody parts.

2.1.2 Based on their application in feed, phytobiotics can be classified into herbs, botanicals, essential oils, and oleoresins (Kollanoor Johny & Venkitanarayanan, 2017)

2.1.2.1 Herbs: Herbs are plants characterized by their flowering nature, non-woody structure, and non-persistent life cycle. These plants typically do not develop a persistent stem above ground and are known for their relatively softer and more flexible structures compared to woody plants.

2.1.2.2 Botanicals: Botanicals refer to either the entire or processed parts of a plant, including roots, leaves, and bark. These plant-derived substances are often utilized for various purposes, such as medicinal, culinary, or cosmetic applications.

2.1.2.3 Essential oils: Essential oils are hydrodistilled extracts derived from the volatile compounds of plants. These oils capture the aromatic essence of the plant.

2.1.2.4 Oleoresins: Oleoresins are extracts derived from plants using nonaqueous solvents. These substances typically contain a mixture of essential oils and resinous components, capturing the concentrated flavors and aromatic properties of the plant. Oleoresins find application in the food and flavor industry, often serving as natural flavoring agents.

3. Extraction

The selection of an appropriate extraction process is vital in herbal processing to ensure optimal extraction of desired active ingredients. The primary advantage lies in methods that efficiently extract plant compounds in a short time with minimal solvent usage. This not only reduces electricity consumption but also mitigates potential degradation of active components, making short extraction times highly desirable in herbal processing (Mohammad Azmin *et al.*, 2016). Some of the most common extraction methods used are :

3.1 Supercritical fluid extraction (SFE)

In this method supercritical fluids (viz. , such CO2) are employed to extract phytochemicals under high pressure and temperature conditions.

- Advantage: The solvent (CO2) is inexpensive and can be recycled easily. Also, pure extraction is possible with this method.
- Limitation: Specialized equipment is required also, there is a chance of losing desired compounds with inappropriate solvent selection

3.2 Microwave Assisted extraction (MAE)

Microwaves are employed to heat the solvent and plant material, enhancing the extraction efficiency by promoting the release of phytochemicals.

- Advantage: this method can be used for both industrial and laboratory scales and can be performed in less time with good profits on capital investment.
- Limitation: this method is not effective in the extraction of nonpolar compounds and cannot be used for heat-sensitive compounds. The equipment used in this process is expensive and difficult to operate.

3.3 Sonication

This method is similar to MAE processes, with microwave power substituted by ultrasonic equipment to enhance the extraction processes.

- Advantage: This is an efficient, quick, easy method with reduced solvent consumption for large-scale commercial applications
- Limitation: not effective for extraction of oil.

3.4 Soxhlet extraction

Soxhlet extraction functions not just as a method for extracting phytochemicals but also serves as a benchmark for evaluating newer extraction techniques. The system operates continuously, with the solvent recirculating through the sample.

- Advantage: It is a very simple and cost-effective method with the added benefit of temperature control, expanding its range of applications.
- Limitation: it is a time-consuming process and also requires a large amount of extractants (solvent).

3.5 Hydrodistillation

The hydrodistillation process closely resembles the Soxhlet extraction process, with the distinction that water serves as the solvent. These two methodologies serve as popular choices for isolating volatile and non-volatile polar components from herbs, showcasing nuances in their solvent selection and equipment configurations.

- Advantage: Due to the elimination of organic solvents in extraction, this method substantially decreases material costs linked with the extraction process.
- Limitation: Separating water from the product takes a long time and requires a lot of energy.

3.6 Steamdistillation extraction

In steam distillation, a dedicated steam generator infuses a plant-solvent mixture with steam, triggering the evaporation of both water and desired fragrant compounds at lower temperatures. Condensation of this vapor mixture then allows for the efficient separation of the solvent from the extracted compounds. This method is commonly used for extracting essential oils, and it is effective in preventing the decomposition of organic compounds.

- Advantage: This method can extract water-insoluble compounds quickly.
- Limitation: extraction might contain impurities.

3.7 Accelerated solvent extraction (ASE)

ASE uses high temperatures (50 to 200°C) and pressures (10 to 15 MPa) to keep the solvent in a liquid state, improve solvent diffusivity, and expedite the extraction process.

- Advantage: This innovative technique stands out as a viable alternative to SFE for extracting polar compounds, delivering a more environmentally friendly approach with lower solvent utilization and quicker extraction kinetics

- Limitation: Has specific applicability for thermally stable extracts.

3.8 Hot water extraction

Hot water extraction (HWE) is a similar technique as the ASE process. Instead of using an organic solvent, this method employs hot water to extract the desired compounds.

- Advantage: It is a low-energy and investment-demanding method effectively used for the extraction of essential oil with high-oxygenated components
- Limitation: More suitable method for heat-resistant compounds.

4. Mode of action

Plants lack an active immune system but have evolved diverse secondary metabolites (SM) over 400 million years to defend against herbivores and microbes (Wink, 2015). These SMs serve various functions, such as protection, attraction, and medicinal properties. These SM are known to human for their beneficial action and have been used for ages in traditional medicine. Herbs are known for their toxic or hallucinogenic properties, often containing alkaloids, terpenoids, or other secondary metabolites targeting molecular elements in animals or humans. These SMs, with specificities for neuroreceptors, enzymes, ion channels, ion pumps, or cytoskeletal elements, are extracted from plants and can be used in feed or as a medicine. Herbs extract may contain a variety of SMs, and the activity of an extract may result from synergistic interactions among several SMs present. These extracts are often reported to treat a broad spectrum of health disorders rather than a single condition. (Asimi & Sahu, 2013; Wink, 2015). As SMs contribute to the therapeutic properties of herbs, it is essential to understand their mechanisms of action to harness their potential for promoting health and treating various medical conditions. SMs can exhibit diverse modes of biological activity, which include:

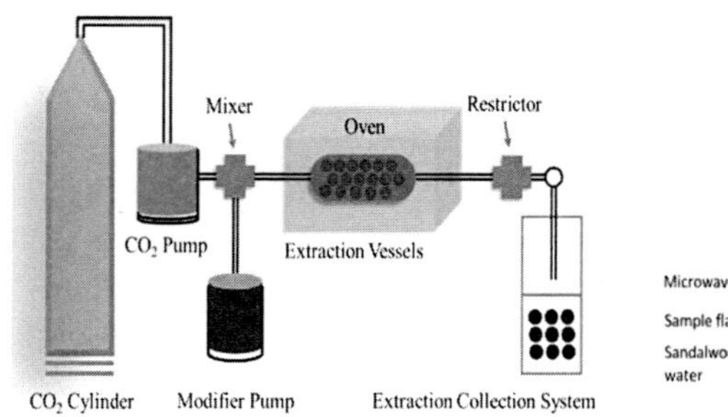

Supercritical fluid extraction

Microwave Assisted extraction

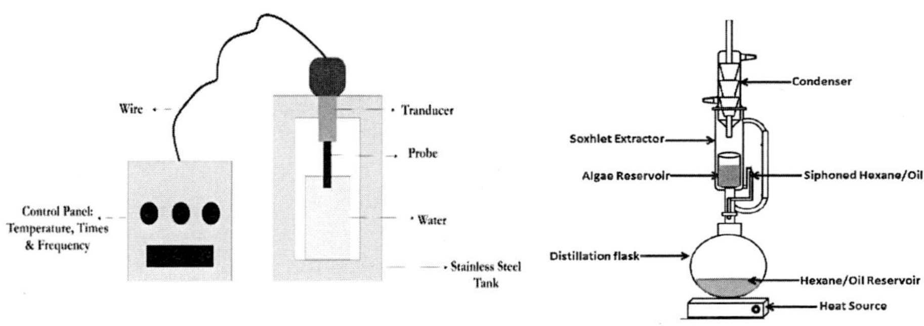

Sonication

Soxhlet extraction

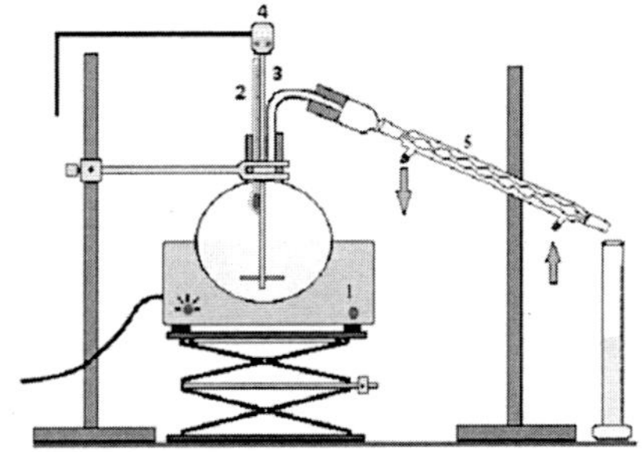

Hydrodistillation 1: Water bath, 2: thermometer, 3: stirrer, 4: Electric motor and 5: condenser

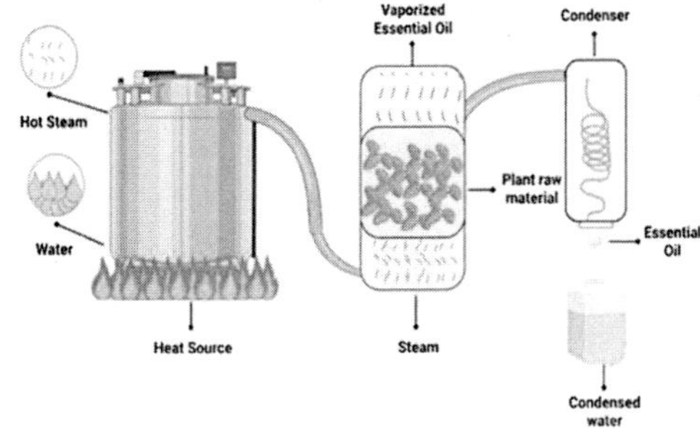

Steamdistillation extraction

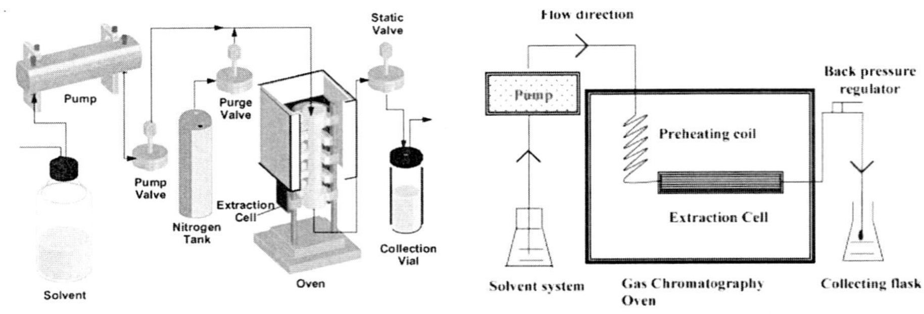

Accelerated solvent extraction Hot water extraction

Fig. 1. Common methods used in extraction of phytochemicals from herbs.

4.1 Covalent modification of protein

In plants, secondary metabolites not only form the structural basis of various compounds but also incorporate reactive groups that can covalently bind to proteins, peptides, and DNA. These modifications, occurring under physiological conditions, can impact protein structure and function. Reactive secondary metabolites may act as "multitarget drugs," influencing various cellular processes. However, their non-selective interactions with proteins can lead to unintended effects (Wink & Schimmer, 2010). Alkylation of DNA bases by these metabolites may result in mutations, potentially causing diseases like cancer. Although some of these compounds have been used in traditional medicine, their use is considered outdated in modern phytotherapy due to associated risks (Wink, 2008).

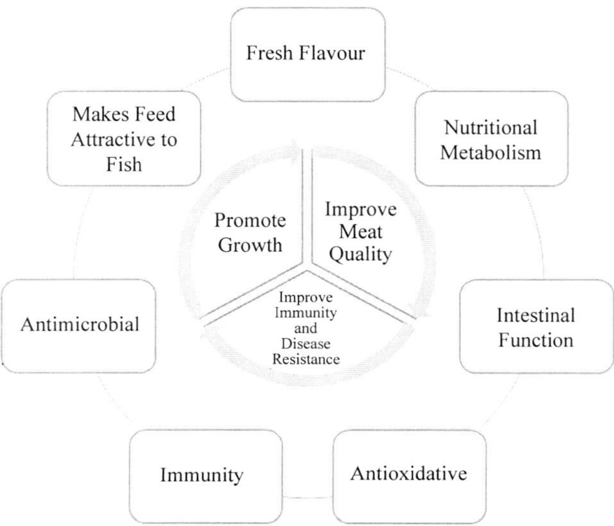

Fig. 2. Roles and mechanisms of herbs in aquaculture application (based on Pu *et al.*, 2017)

4.2 Non-covalent modification of protein

Proteins in cells are affected not only by secondary metabolites (SMs) with reactive groups but also by phenols and polyphenols found in herbal medicines. Phenolic compounds, common in these medicines, form hydrogen bonds with proteins and create ionic bonds with amino groups, impairing protein flexibility like SMs with reactive groups. Glycosylated phenolics enhance interactions through hydrogen bonding (Wink, 2015). Polyphenols act as multitarget drugs, influencing proteins and transcription factors impacting gene regulation. Transcriptome analyses reveal the broad effects of phenolic compounds on proteins and genes, emphasizing their pleiotropic nature.

4.3 Interaction with membrane

Living organisms have protective biomembranes preventing uncontrolled substance movement. Biomembranes, with proteins for communication, can be influenced by certain lipophilic secondary metabolites (SM) (Wink, 2015). Lipophilic SM-like terpenes in essential oils alter membrane properties, increasing permeability and showing antimicrobial effects. Mint oil, for instance, affects calcium channels in intestinal muscle cells (Wyk & Wink, 2015).

4.4 Interaction with nucleic acid

Certain natural secondary metabolites, have the ability to interact with DNA, which can lead to changes in its structure and potentially increase the risk of cancer. Despite this risk, plants containing these compounds have been used in traditional medicine worldwide because they possess strong antibacterial, antifungal, antiviral, and cell-killing properties. Some examples of these compounds include pyrrolizidine alkaloids found in plants of the Boraginaceae family, aristolochic acids in Aristolochia plants, cycasine in cycads, furanocoumarins in plants of the Apiaceae family, and ptaquiloside in bracken ferns (Pteridium aquilinum).

4.5 Antioxidant Properties

Excess reactive oxygen species (ROS) can harm vital cell components, contributing to chronic conditions like metabolic syndrome, cardiovascular diseases and potentially cancer. Certain compounds in plants, such as phenolics, terpenoids, and ascorbic acid, have antioxidant properties, countering ROS effects. Many herbal remedies and algae-derived products rich in these compounds may act as antioxidants, along with influencing proteins and biomembranes (Wink, 2015).

Table 1. Some commonly used herbal plant extracts in aquaculture

English name	Scientific name	Useful part	Therapeutic properties	Reference
Nutmeg	Myristica flagrans	Seed	Digestion stimulant, antidiarrhoeal, enhanced growth and immunity, reduced stress	(Rashidian *et al.*, 2022)
Cinnamon	Cinnamomum sp.	Bark, leaf	Reduces lipid content, improves flesh flavour and texture	(Dedi *et al.*, 2016)
Clove	Syzygium aromaticum	Cloves	antioxidant and stimulant for fish growth	(Gaber, 2000)
Black Cardamom	Amomum subulatum	Seed	Improves growth potential, antioxidant status, immune parameters	(Ali *et al.*, 2024)
Ginger	Zingiber officinale Roscoe	Rhizome	increase in growth, feed conversion, protein efficiency, bactericidal and anti-protease	(Nya & Austin, 2009)
Garlic	Allium sativum	Bulb	growth promotion, antimicrobial, antiviral, antioxidant, and antiparasitic	(Valenzuela-Gutiérrez *et al.*, 2021)
Turmeric	Curcuma longa	Rhizome	Antioxidant, anti-inflammatory, improve growth	(Fagnon *et al.*, 2020)
Thyme	Thymus ulgaris	Whole plant	Digestion stimulants, enhancing of growth, immune responses, and disease resistance.	(Zargar *et al.*, 2019)
Peppermint	Mentha piperita	Leaves	improved survival, weight gain, Appetite and digestion stimulant, anti-protease and bactericidal activities	(Talpur, 2014)
Neem	Azadirachta indica	Leaves, bark	Improve immunity, growth Antiviral, antiseptic, fungicidal	(Talpur & Ikhwanuddin, 2013)

5. Advantages of use of herbs in aquaculture

5.1 Enhanced feed acceptance: Herbs can improve the palatability of fish feed, making it more appealing to the fish and improving feed acceptance.

5.2 Natural Immunostimulants: Certain herbs are known for their immunostimulant properties that can boost the immune system of fish, enhancing their resistance to diseases and thus enhancing production.

5.3 Antioxidant Properties: Herbs are rich sources of antioxidants, which can help reduce oxidative stress in intensive fish farming, supporting overall health and productivity.

5.4 Improved Growth Performance: Herbs can promote efficient nutrient utilization and metabolism, thereby contributing to better growth performance in fish.

5.5 Disease Prevention: Herbs with antimicrobial and antiparasitic properties can effectively prevent and control diseases in aquaculture.

5.6 Sustainable Aquaculture: herbs are a sustainable alternative to synthetic additives and chemicals, supporting the increasing demand for eco-friendly aquaculture practices. Also plays a crucial role in organic aquaculture.

5.7 Bioactive Compounds: Herbs contain bioactive compounds that can positively influence various physiological functions in fish, including digestion, nutrient absorption, and maturation breeding.

6. Challenges in the application of herbs in aquaculture

Despite of numerus advantages of herbs is not devoid of problems. The challenges facing the extensive use of herbs and herbal extracts in aquaculture can be outlined as follows:

6.1 Complex Composition: Herbs contain numerous effective ingredients, making it challenging to identify and understand them all. The bioavailability of these ingredients is influenced by factors like growth stage and cultivation location, further complicating their precise identification and dosage determination.

6.2 Commercial Application: Applying herbs and herbal extracts to feed or water under commercial conditions poses difficulties. Refining herbal products and extracting effective ingredients economically

6.3 Lack of Functional Mechanistic Understanding: There is a scarcity of studies investigating the functional mechanisms of herbs and its extracts. Research has predominantly focused on fish production and disease management, leaving the specific mechanisms of action for most herbs unclear. Traditional knowledge of herbs, developed for humans, may not fully explain how herbs function in aquatic animals with distinct metabolic and immunization mechanisms.

Addressing these challenges requires further research to control the quality of medicinal herbs, refine application methods, and gain a deeper understanding of the functional mechanisms involved in aquaculture.

Conclusion

In conclusion, incorporating herbs into fish feed as functional additives emerges as a promising strategy to enhance aquaculture practices. These herbs have demonstrated potential not only in stress reduction, immunity enhancement, and bacterial control but also in promoting growth, food intake, and assimilation. Nevertheless, challenges such as determining precise dosages, variability in herbal composition, and a limited understanding of specific mechanisms of action need to be addressed. Despite these obstacles, the shift towards herbal supplementation in fish feed resonates with the increasing demand for sustainable and natural alternatives in aquaculture. Further research and standardization efforts are needed to unlock the full potential of herbs, ensuring their consistent and effective application in fish nutrition and health management.

References

Ali, W., Fatima, M., Shah, S. Z. H., Khan, N., & Naveed, S. (2024). Black cardamom (Amomum subulatum) extract improves growth potential, antioxidant status, immune parameters and response to crowding stress in Catla catla. Journal of Animal Physiology and Animal Nutrition, 108(1), 274–284. https://doi.org/10.1111/jpn.13888

Asimi, O., & Sahu, N. (2013). Herbs/spices as feed additive in aquaculture. Scientific Journal of Pure and Applied Sciences, 2. https://doi.org/10.14196/sjpas.v2i8.868

Bondad Reantaso, M. G., MacKinnon, B., Karunasagar, I., Fridman, S., Alday, Sanz, V., Brun, E., Le Groumellec, M., Li, A., Surachetpong, W., Karunasagar, I., Hao, B., Dall'Occo, A., Urbani, R., & Caputo, A. (2023). Review of alternatives to antibiotic use in aquaculture. Reviews in Aquaculture, 15(4), 1421–1451. https://doi.org/10.1111/raq.12786

Chang, J. (2000). Medicinal herbs: Drugs or dietary supplements? Biochemical Pharmacology, 59(3), 211–219. https://doi.org/10.1016/S0006-2952(99)00243-9

Dedi, J., Hutama, A. A., Nurhayati, T., Vinasyiam, A., & others. (2016). Growth performance and flesh quality of common carp, Cyprinus carpio feeding on the diet supplemented with cinnamon (Cinnamomum burmannii) leaf. Aquaculture, Aquarium, Conservation & Legislation, 9(5), 937–943.

Fagnon, M. S., Thorin, C., & Calvez, S. (2020). Meta-analysis of dietary supplementation effect of turmeric and curcumin on growth performance in fish. Reviews in Aquaculture, 12(4), 2268–2283. https://doi.org/10.1111/raq.12433

Gaber, M. (2000). Growth Response of Nile Tilapia Fingerlings (Oreochromis Niloticus) Fed Diets Containing Different Levels of Clove Oil. Egyptian Journal of Aquatic Biology and Fisheries, 4(1), 1–18. https://doi.org/10.21608/ejabf.2000.1637

Kollanoor Johny, A., & Venkitanarayanan, K. (2017). Preharvest Food Safety—Potential Use of Plant-Derived Compounds in Layer Chickens. In Producing Safe Eggs (pp. 347–372). Elsevier. https://doi.org/10.1016/B978-0-12-802582-6.00017-3

Mohammad Azmin, S. N. H., Abdul Manan, Z., Wan Alwi, S. R., Chua, L. S., Mustaffa, A. A., & Yunus, N. A. (2016). Herbal Processing and Extraction Technologies. Separation & Purification Reviews, 45(4), 305–320. https://doi.org/10.1080/15422119.2016.1145395

Nya, E. J., & Austin, B. (2009). Use of dietary ginger, Zingiber officinale Roscoe, as an immunostimulant to control Aeromonas hydrophila infections in rainbow trout, Oncorhynchus mykiss (Walbaum). Journal of Fish Diseases, 32(11), 971–977. https://doi.org/10.1111/j.1365-2761.2009.01101.x

Pu, H., Li, X., Du, Q., Cui, H., & Xu, Y. (2017). Research progress in the application of Chinese herbal medicines in aquaculture: A review. Engineering, 3(5), 731–737. https://doi.org/10.1016/J.ENG.2017.03.017

Rashidian, G., Shahin, K., Elshopakey, G. E., Mahboub, H. H., Fahim, A., Elabd, H., Prokić, M. D., & Faggio, C. (2022). The Dietary Effects of Nutmeg (Myristica fragrans) Extract on Growth, Hematological Parameters, Immunity, Antioxidant Status, and Disease Resistance of Common Carp (Cyprinus carpio) against Aeromonas hydrophila. Journal of Marine Science and Engineering, 10(3), 325. https://doi.org/10.3390/jmse10030325

Syahidah, A., Saad, C. R., Daud, H. M., & Abdelhadi, Y. M. (2015). Status and potential of herbal applications in aquaculture: A. review. https://aquadocs.org/bitstream/handle/1834/11830/IFRO-v14n1p27-en.pdf?sequence=1

Talpur, A. D. (2014). Mentha piperita (Peppermint) as feed additive enhanced growth performance, survival, immune response and disease resistance of Asian seabass, Lates calcarifer (Bloch) against Vibrio harveyi infection. Aquaculture, 420–421, 71–78. https://doi.org/10.1016/j.aquaculture.2013.10.039

Talpur, A. D., & Ikhwanuddin, M. (2013). Azadirachta indica (neem) leaf dietary effects on the immunity response and disease resistance of Asian seabass, Lates calcarifer challenged with Vibrio harveyi. Fish & Shellfish Immunology, 34(1), 254–264. https://doi.org/10.1016/j.fsi.2012.11.003

Valenzuela-Gutiérrez, R., Lago-Lestón, A., Vargas-Albores, F., Cicala, F., & Martínez-Porchas, M. (2021). Exploring the garlic (Allium sativum) properties for fish aquaculture. Fish Physiology and Biochemistry, 47(4), 1179–1198. https://doi.org/10.1007/s10695-021-00952-7

Valladão, G. M. R., Gallani, S. U., & Pilarski, F. (2015). Phytotherapy as an alternative for treating fish disease. Journal of Veterinary Pharmacology and Therapeutics, 38(5), 417–428. https://doi.org/10.1111/jvp.12202

Wink, M. (2008). Evolutionary Advantage and Molecular Modes of Action of Multi-Component Mixtures Used in Phytomedicine. Current Drug Metabolism, 9(10), 996–1009. https://doi.org/10.2174/138920008786927794

Wink, M. (2015). Modes of Action of Herbal Medicines and Plant Secondary Metabolites. Medicines, 2(3), 251–286. https://doi.org/10.3390/medicines2030251

Wink, M., & Schimmer, O. (2010). Molecular Modes of Action of Defensive Secondary Metabolites. In M. Wink (Ed.), Functions and Biotechnology of Plant Secondary Metabolites (pp. 21–161). Wiley-Blackwell. https://doi.org/10.1002/9781444318876.ch2

Wyk, B.-E. van, & Wink, M. (Eds.). (2015). Phytomedicines, Herbal Drugs, and Poisons. University of Chicago Press. https://press.uchicago.edu/ucp/books/book/chicago/P/bo19196930.html

Zargar, A., Rahimi, Afzal, Z., Soltani, E., Taheri Mirghaed, A., Ebrahimzadeh-Mousavi, H. A., Soltani, M., & Yuosefi, P. (2019). Growth performance, immune response and disease resistance of rainbow trout (Oncorhynchus mykiss) fed Thymus vulgaris essential oils. Aquaculture Research, 50(11), 3097–3106. https://doi.org/10.1111/are.14243

10

Application of Aquatic Algae for Indication and Controlling of Pesticide Pollution

Muskan Kosriy[1*], Anand Vaishnav[1], Deepika Korram[1], Priyanka Acharya[1] and Varsha Sahu[1]

[1]*Late Shri Punaram Nishad College of Fisheries, Kawardha, DSVCKV Anjora, Durg-491995, Chhattisgarh, India*

[2]*College of Fisheries, Central Agricultural University (Imphal) Lembucherra, 799210, Tripura India*

Abstract

Aquatic algae play a crucial role as bioindicators of water quality, particularly in the context of pesticide pollution. Pesticides are essential for enhancing agricultural productivity, and pose environmental and health risks when mismanaged. This chapter explores the historical development of algal biomonitoring, emphasizing its global recognition and diverse applications in contemporary ecological assessments. Algae are sensitive to pollutants and serve as early indicators of water quality deterioration, with specific forms identified as markers of contamination. Microalgae play a crucial role in pesticide removal through various mechanisms, including bioadsorption, bioaccumulation, and biodegradation processes. highlighted their versatility in actively mitigating pesticide contamination. Pioneering studies utilizing benthic microalgae for sediment remediation have underscored the practical applications of algae for addressing pesticide challenges. The potential use of certain algae as metal-specific bioindicators provides insight into their role in assessing metal pollution. Microalgae-based bioremediation strategies, particularly phytoremediation, have emerged as environmentally friendly and cost-effective solutions for pesticide contamination. Recycling microalgae after wastewater treatment yields additional benefits including biodiesel and biochar production. However, challenges such as large-scale industrial applications, efficient harvesting, pollutant absorption, and genetic modification warrant further exploration.

***Keyword**: Biochar, Bioadsorption, Bioaccumulation, Microalgae*

1. Introduction

Aquatic algae is present in both contaminated and uncontaminated water sources. serve as essential bioindicators for evaluating water quality. Their sensitivity to pollutants and position in The significance of lower-level organisms in the food chain lies in their invaluable role in comprehending the effects of toxic substances on aquatic ecosystems. (Dwivedi & Pandey (2002). Pesticides, including herbicides, fungicides, and insecticides, play a crucial role in agriculture by utilizing forestry and animal husbandry to improve crop production. However, their indiscriminate use can lead to environmental contamination, which poses significant risks to ecosystems and human health. Pesticide contamination primarily results from surface runoff, emissions during spraying, improper transportation, inadequate spraying practices, and insufficient wastewater treatment from the pesticide industry. Introduction of pesticides into the environment Presents numerous environmental and health risks., inhibiting plant and animal growth and affecting the development and nervous systems of animals Pesticides have the potential to infiltrate the food chain, bioaccumulate in higher trophic levels, and reach the human body through biomagnification. Commonly used pesticides, such as imidacloprid, isoproturon, simazine, atrazine, mecoprop, and glyphosate, along with historical organochlorine pesticides (OCPs), contribute to significant pesticide contamination.

Considering the risks posed by pesticides to human health and the environment, effective remediation strategies are essential. Bioremediation technology, particularly that using microalgae, has gained attention because of its environmentally friendly characteristics, cost-effectiveness, and minimized potential for secondary pollution. Microalgae, such as Chlorella and Spirulina, show promise for removing nutrients, pesticides, toxic elements, and other contaminants from wastewater. Their adaptability, wide geographical distribution, and application in various industries make microalgae a valuable tool for environmental restoration.

2. Algae as Bioindicators

Aquatic algae offer several advantages as bioindicators, including their short life cycles, direct responsiveness to environmental factors, cost-effective monitoring, and non-destructive sampling. These attributes enable rapid and efficient responses to changes in their surroundings (Trainor, 1984). The use of algal species presence and absence as indicators has evolved over time. Diatoms, with their diversity in unpolluted zones and specific algae in different zones, provide insights into water quality variations (Hynes, 1969). This concept has been expanded to encompass population and biodiversity indicators, recognizing the impact of environmental disturbances on biological systems. specific algae P. caudate, Euglena sp., C. globosa, S. limorphus, Ulothrix sp., O. capilliforme, Spirogyra sp., N. cuspidate, N. palea, and BGA, are identified as indicators of contaminated water bodies. The dominance of specific algae, especially BGA, serves as an early indicator of deteriorating water quality (Endmondson 1991). Metal concentration data from high biomass-producing algae in polluted water bodies revealed variations in metal uptake among different species. Cyanophycean granules in certain algae such as O. nigra and P. bohneri suggest their potential use as metal-specific bioindicators, pending further validation under varying ecological conditions. Algae, especially diatoms, have a rich history in Europe's biomonitoring practices and are integral to international initiatives, such as the Water Framework Directive (WFD) of the European Union. Countries worldwide, including New Zealand, South Africa, Canada, Israel, Brazil, and Australia, have utilized algae-centric monitoring protocols, emphasizing their global recognition (Barinova *et al.*, 2006; Lobo *et al.*, 2004; Chessman *et al.*, 2007).

3. Algae in Contemporary Ecological Assessments

Contemporary ecological assessments involving algae aim to understand the causes and effects of proliferation, accumulation, and demise of various algal species. These assessments are crucial for ensuring safe drinking water and productive fisheries, and employ remnants of diatoms and chrysophytes in lake sediments to deduce impacts such as acid rain and climate change (Charles *et al.*, 1990; Smol *et al.*, 2005). Algal community composition in different aquatic habitats serves as a regulatory tool in many countries to ensure water quality standards (Passy and Bode, 2000; Passy and Bode, 2004).

4. Importance of Algae in Monitoring Pesticide Pollution

The introduction of organic pollutants, particularly pesticides, poses a chronic threat to human health and wildlife. The persistence and accumulation of these chemicals in sediments necessitate effective monitoring and control strategies. Algae, which are sensitive to environmental changes, play a vital role in the assessment and mitigation of pesticide pollution. Despite the pollution, water bodies support a significant diversity of algae, with dominant forms varying seasonally. Changes in algal communities were observed and linked to severe winter conditions, causing mass mortality of unicellular forms and heavy rainfall during the rainy season, leading to the dilution of algae in the water system. The contamination of water bodies through point sources was primarily marked by the dominance of blue-green algae (BGA), specifically O. nigra and P. bohneri, which remained year-round. In particular, O. nigra is known for its cost-effective contribution to ecotechnology by assisting in the elimination of chromium (Cr) from effluents produced by the tannery industry (Rai *et al.*, 2005). Alterations in community composition due to the physicochemical characteristics of water bodies serve as an early indicator of deteriorating water quality. Species with stringent environmental requirements degrade swiftly in the presence of pollutants, rendering them reliable indicators of pristine ecosystems. On the other hand, some types of algae such as P. caudate, Euglena sp., C. globosa, S. limorphus, Ulothrix sp., O. capilliforme, Spirogyra sp., N. cuspidate, N. palea, and BGA were recognised as indications of water bodies that are contaminated. The prevalence of BGA, particularly Oscillatoria, indicates severe eutrophication in highly contaminated water bodies.

Metal concentration data from high biomass-producing algae in polluted water bodies revealed variations in metal uptake depending on the species, developmental stage, and the presence of different metallic contaminants. BGA is a non-heterocystous species that prevails in highly contaminated water bodies, likely owing to inherent defensive mechanisms and external surface sorption mechanisms against toxic heavy metals. The cyanophycean granules in BGA cells may serve as nitrogen storage, accumulating high concentrations of metals under environmental stresses and constituting a component of the internal detoxification system (Fernandez-Pinas *et al.* 1995; Scott and Palmer 1990; Thomas 1993), the relationship between metal accumulation and background metal levels exhibited variability across different algal forms, with some showing positive correlations and others showing negative correlations with ambient metal concentrations. Notably, O. nigra and P. bohneri have been identified as potential metal accumulators, suggesting their potential use in the development of metal-specific bioindicators. However, further research is required to validate the effectiveness of these indicator organisms under different ecological conditions with varying levels of metal contamination.

Table 1. Management of microalgae for control of pesticide

S. No	Aspect	Details
1.	Cultivation Methods	Microalgae cultivation can be achieved through three methods: **1. Photoautotrophic:** Utilizes carbon dioxide (CO2), light, and nutrients for photosynthesis. Suitable for areas near abundant CO2 sources. **2. Heterotrophic:** Uses organic compounds as both energy and carbon sources, overcoming light limitations. **3. Mixotrophic:** A combination of photoautotrophic and heterotrophic processes. Allows the use of both organic and inorganic carbon, overcoming light limitations.
2.	Review and Recognition	Microalgae can be screened and identified morphologically and molecularly. Morphological classification involves color, shape, texture, and cell arrangement. Molecular identification uses techniques like 18S rDNA to identify specific algal strains. Advanced microscopy, such as TEM and SEM, is used for characterization.
3.	Management Measures for Pesticides	Various pesticides have specific management measures implemented by regulatory authorities. For example, Endosulfan, bromomethane, acephate, carbosulfan, dimethoate, butyl 2,4-dichlorophenoxyacetate, paraquat, dicofol, flubendiamide, carbofuran, phorate, isofenphos-methyl, aluminium phosphide, methidathion E.C, chloropicrin have specific regulations and prohibition dates.
4.	Metabolic Pathways Involved in Pesticide Elimination by Microalgae	Microalgae employ various mechanisms for pesticide removal: **1. Bioadsorption:** Passive process involving the adsorption of pesticides onto microalgae surfaces through mechanisms Including electrostatic interaction, surface complexation, ion exchange, absorption, and precipitation, these processes are influenced by the presence of surface-active groups and the inherent properties of microalgae. **2. Bioaccumulation:** Active process where pesticides gradually accumulate in microalgae biomass. **3. Biodegradation:** Microalgae metabolize and break down pesticides into less harmful compounds.

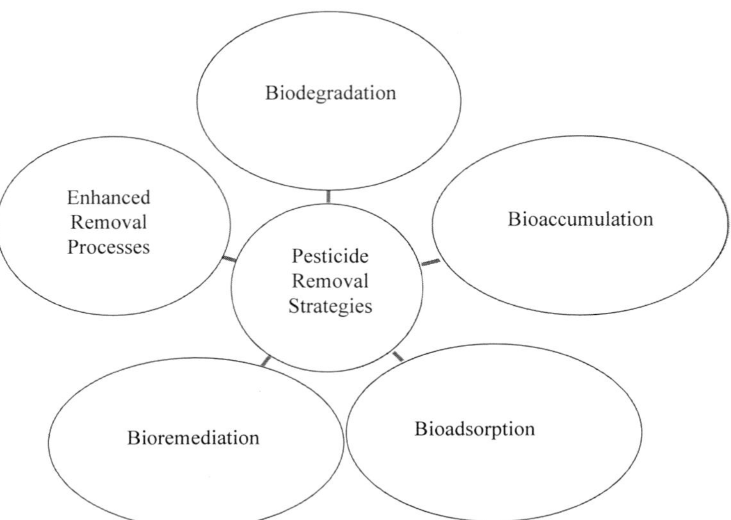

Fig. 1. Pesticide removal strategies

4.1 Bioadsorption: Unraveling Passive Pesticide Sequestration

Bioadsorption, a process integral to passive pesticide sequestration, is a promising environmental remediation mechanism. This method involves the natural capacity of living organisms, such as bacteria, fungi, algae, or plant materials, to adsorb and accumulate pesticides from their surroundings. As pesticides increasingly pose threats to ecosystems and water bodies, understanding and harnessing bioadsorption offers an innovative approach for mitigating their impact. The affinity of biological entities for pesticides not only provides a means of removal but also holds potential for developing eco-friendly and sustainable remediation strategies. By unraveling the intricacies of passive pesticide sequestration through bioadsorption, we can pave the way for effective and environmentally responsible solutions for pollution control and ecosystem preservation.

4.2 Bioaccumulation: Orchestrating Active Pesticide Uptake

Bioaccumulation, a dynamic process inherent to microalgae, is intricately orchestrated through various biochemical mechanisms. The bioconcentration factor (BCF) plays a pivotal role in quantifying the concentration ratio of pesticides within microalgae in relation to their surrounding environment. When microalgae encounter pesticides, a sophisticated defense mechanism comes into play. Exposure triggers the production of reactive oxygen species (ROS), prompting the activation of antioxidant enzymes such as superoxide dismutase (SOD), catalase (CAT), and ascorbate peroxidase (APX). This coordinated response not only safeguards microalgae from oxidative stress but also facilitates the active uptake and biodegradation of pesticides. The synergy between bioaccumulation and detoxification processes underscores the adaptability of microalgae to environmental challenges, offering insights into the intricate dynamics of pesticide interactions within aquatic ecosystems.

4.3 Biodegradation: Microalgae as Microbial Pesticide Detoxifiers

Playing a pivotal role in microbial pesticide removal, microalgae participate in the essential process of biodegradation. This intricate mechanism involves a cascade of enzymatic activities, including esterase, transferase, cytochrome P450, hydrolase, phosphatase, phosphotriesterase, oxygenase, and oxidoreductases. The biodegradation process includes the activation of pesticides by cytochrome P450 and conjugation with glutathione, glucose, and malonate, followed by the transport of conjugates into vacuoles.

4.4 Enhanced Removal Processes: Maximizing Microalgal Potential

The strategic acclimation of microalgae strains to extreme wastewater forms the basis for enhanced removal processes. Co-cultivation studies with bacteria have revealed synergistic or antagonistic effects on pollutant removal, particularly the promising interactions between cyanobacteria and bacteria. Additionally, immobilization techniques, such as surface adsorption, covalent bonding, crosslinking, and embedding, augment the performance of microalgae by preserving their catalytic activity, facilitating easy separation, and enhancing their survival in challenging environments. These comprehensive strategies underscore the substantial potential of microalgae to address pesticide contamination in water systems.

4.5 Bioremediation

Growing interest in bioremediation, specifically phytoremediation, has highlighted the potential of algae to remove or detoxify environmental pollutants. The investigation of the bioaccumulation and biodegradation of organic xenobiotics in green algae contributes significantly to environmental remediation efforts (Suresh and Ravishankar, 2004). In particular, microalgae have demonstrated capabilities in the bioremediation of colored wastewater, emphasizing their role in environmental sustainability (Huang *et al.*, 2010; Ellis *et al.*, 2012; Lim *et al.*, 2010). In a pioneering study, benthic microalgae were successfully employed for remediation of organically enriched sediments (Yamamoto *et al.*, 2008). This exemplifies the potential of algae, especially microalgae, in addressing the challenges posed by pesticide pollution in aquatic ecosystems

5. Eco-Friendly Harnessing of Microalgae in Wastewater Treatment

5.1 Biodiesel Production

Third-generation biodiesel sourced from microalgae has emerged as a promising alternative to traditional fossil fuels, boasting renewability, degradability, environmental friendliness, and lower greenhouse gas emissions (Ahmad *et al.*, 2016; Ahmad *et al.*, 2011). In contrast to first- and second-generation raw materials from plants, microalgae cultivation requires minimal agricultural land, alleviating concerns related to pesticide pollution and excessive fertilizer use (Aghbashlo and Demirbas, 2016). Microalgae cultivated in wastewater present a dual advantage of both degrading pollutants and yielding biomass, thereby reducing the economic costs associated with farmland restoration (Smith *et al.*, 2010).

Maximizing biodiesel production from microalgae involves leveraging nonagricultural land and utilizing CO_2 as a carbon source, as highlighted by Zhou *et al.*, (2013). Essential considerations in this process include optimizing culture conditions, enhancing lipid content, and employing transesterification methods for oil production, as detailed by Chen *et al.*, (2011), Suganya *et al.*, (2016) and Zhu *et al.*, (2017). Nutrient limitations, such as nitrogen, phosphorus, and oxygen scarcity, play a crucial role in boosting oil production rates (Zhu, 2014). The selection of harvesting methods, including centrifugation, ultrafiltration, sedimentation, air flotation, and flocculation, is contingent on the specific cultivation methods employed, as outlined by Schenk *et al.*, (2008) and Zhu *et al.*, (2017). To achieve biodiesel conversion, microalgae oil undergoes a sequence of pyrolysis followed by reverse esterification, as explained by Suganya *et al.*, (2016).

5.2 Biochar Production

Biochar produced through biomass pyrolysis under oxygen-deficient conditions has garnered significant attention for its role in environmental restoration, as highlighted by Sohi (2012) and Song *et al.*, (2018). Owing to its substantial surface area and well-developed micropore structure, biochar is a highly effective adsorbent for water pollution, as indicated by Tan *et al.*, (2015) and Song *et al.*, (2019a). Its versatile applications encompass adsorption for air and water pollution, catalytic functions in syngas cleaning, contributions to biodiesel production, air pollution control, and soil improvement, as elucidated by Ding *et al.*, (2016) and Thomas and Gale (2015). The repurposing of traditional landfilled or incinerated waste biomass into biochar offers a valuable contribution to waste recycling, as discussed by Cao *et al.*, (2019). A novel technique involving microwave-assisted low-temperature hydrothermal treatment for algae hydrochar production is emerging, with shorter reaction times and enhanced carbon recovery, as proposed by Cao *et al.*, (2019). Nonetheless, further research on algal biochar, particularly its adsorption mechanisms, is required (Zheng *et al.*, 2017). This integrated approach highlights the potential of microalgae not only for wastewater treatment but also for generating valuable products such as biodiesel and biochar for sustainable environmental practices.

6. Current Limitations and Future Perspectives

Despite significant progress, microalgal biotechnology still faces challenges.

6.1 Large-scale industrial applications

- High production costs hinder commercial operations, experiments are mainly conducted at medium or photobioreactor levels, and outdoor factors impact reproduction under open conditions.

6.2 Microalgae Harvesting

- Small size and hydrophilic nature pose challenges for gravity settlement.
- Harvesting efficiency and cost-effectiveness issues persist.

6.3 Treatment and Utilization

- Pollutant absorption may lead to toxic substances in the microalgal biomass.
- Proper addressing of accumulated pollutants is crucial for downstream applications.

6.4 Life-cycle analysis (LCA) and Environmental Impact

- Limited application in large-scale industrial settings.
- Current assessments based on laboratory data and theoretical assumptions.
- Comparative studies with soybean biodiesel and petroleum diesel revealed the environmental benefits of algal biodiesel.

6.5 Genetic Modification of Microalgae

- Genetic engineering of pesticide resistance is an emerging field of research.
- Limited research reports on genetically modified microalgae for pesticide removal.

Conclusion

The use of aquatic algae as bioindicators is indispensable for assessing the water quality, particularly in the context of pesticide pollution. Pesticides, which are vital for enhancing agricultural productivity, pose significant environmental and health risks when improperly managed. Algae, which are sensitive to pollutants and play a pivotal role in the food chain, offer a valuable tool for monitoring and mitigating the impacts of pesticide contamination. The historical development of algal biomonitoring, from early indicators pioneered by Kolkwitz and Marsson to contemporary ecological assessments, reflects the global recognition of the importance of algae in safeguarding aquatic ecosystems. Algae, particularly diatoms, serve as key indicators in monitoring programs worldwide, contributing to initiatives such as the Water Framework Directive in the European Union. The dominance and diversity of algal species in response to pesticide pollution highlight their role as early indicators of water quality deterioration. Changes in the community structure, metal accumulation, and specific algal forms serve as crucial markers for assessing the impact of pesticides on aquatic environments. Furthermore, the potential use of certain algae, such as Oscillatoria nigra and P. bohneri, as metal-specific bioindicators holds promise, although further research is required for validation. Bioremediation strategies employing microalgae have emerged as environmentally friendly and cost-effective solutions for pesticide contamination. The mechanisms of bioadsorption, bioaccumulation, and biodegradation demonstrate the versatility of microalgae in actively removing and detoxifying pesticides. Pioneering studies using benthic microalgae for sediment remediation exemplify the practical applications of algae in addressing pesticide challenges.The recycling of microalgae after wastewater treatment presents additional benefits, including biodiesel and biochar production. Microalgae-derived biodiesel offers a promising alternative to traditional fuels with the added advantage of wastewater pollutant removal. Biochar produced from algal biomass demonstrates potential as an effective adsorbent for water pollution, contributing to environmental restoration. Despite the progress in microalgal biotechnology, challenges such as large-scale industrial application, efficient harvesting, pollutant absorption, and genetic modification need further exploration. Life Cycle Analysis (LCA) and environmental impact assessment are crucial for understanding the overall sustainability of microalgae-based remediation strategies. Genetic engineering holds promise for the development of pesticide-resistant microalgae; however, research in this field is still in its infancy.

References

Aghbashlo, M., & Demirbas, A. (2016). Biodiesel: hopes and dreads. Biofuel Research Journal, 3(2), 379-379.

Ahmad, M. N., Mokhtar, M. N., Baharuddin, A. S., Hock, L. S., Ali, S. R. A., Abd-Aziz, S., & Hassan, M. A. (2011). Changes in physicochemical and microbial community during co-composting of oil palm frond with palm oil mill effluent anaerobic sludge. BioResources, 6(4), 4762-4780.

Barinova, S., Smith, T., & Tsarenko, P. (2022). Microalgae, in Spatial Assessment of the Drainage Basin, Influences on the Ecosystem of Lake Agmon, Israel. Applied Microbiology, 2(1), 197-214.

Cao, X., Xi, Y., Liu, J., Chu, Y., Wu, P., Yang, M., ... & Xue, S. (2019). New insights into the CO2-steady and pH-steady cultivations of two microalgae based on continuous online parameter monitoring. Algal Research, 38, 101370.

Charles, D. J., & Simon, J. E. (1990). Comparison of extraction methods for the rapid determination of essential oil content and composition of basil. Journal of the American Society for Horticultural Science, 115(3), 458-462.

Chen, C., Qian, Y., Chen, Q., Tao, C., Li, C., & Li, Y. (2011). Evaluation of pesticide residues in fruits and vegetables from Xiamen, China. Food Control, 22(7), 1114-1120.

Chessman, B., Williams, S., & Besley, C. (2007). Bioassessment of streams with macroinvertebrates: effect of sampled habitat and taxonomic resolution. Journal of the North American Benthological Society, 26(3), 546-565.

Ding, A. J., Huang, X., Nie, W., Sun, J. N., Kerminen, V. M., Petäjä, T., ... & Fu, C. B. (2016). Enhanced haze pollution by black carbon in megacities in China. Geophysical Research Letters, 43(6), 2873-2879.

Dwivedi, B.K. and Pandey, G.C. 2002. Physicochemical factors and algal diversity of two ponds (GirijaKund and Maqubara Pond), Faizabad, India.Pollution Research 21(3): 361-369

Edmondson, D. G., Brennan, T. J., & Olson, E. N. (1991). Mitogenic repression of myogenin autoregulation. Journal of Biological Chemistry, 266(32), 21343-21346.

Hill, P. S., Milligan, T. G., & Geyer, W. R. (2000). Controls on effective settling velocity of suspended sediment in the Eel River flood plume. Continental shelf research, 20(16), 2095-2111.

Hynes, H. B. N. (1969). The enrichment of streams.

Lobo, E. A., Callegaro, V. L. M., Gómez, N., & Ector, L. (2004). Review of the use of microalgae in South America for monitoring rivers, with special reference to diatoms. Vie et Milieu/ Life & Environment, 105-114.

Passy, S. I., & Bode, R. W. (2004). Diatom model affinity (DMA), a new index for water quality assessment. Hydrobiologia, 524, 241-252.

Rai, H. S., Bhattacharyya, M. S., Singh, J., Bansal, T. K., Vats, P., & Banerjee, U. C. (2005). Removal of dyes from the effluent of textile and dyestuff manufacturing industry: a review of emerging techniques with reference to biological treatment. Critical reviews in environmental science and technology, 35(3), 219-238.

Schenk, P. M., Thomas-Hall, S. R., Stephens, E., Marx, U. C., Mussgnug, J. H., Posten, C., ... & Hankamer, B. (2008). Second generation biofuels: high-efficiency microalgae for biodiesel production. Bioenergy research, 1, 20-43.

Smith, R., Middlebrook, R., Turner, R., Huggins, R., Vardy, S., & Warne, M. (2012). Large-scale pesticide monitoring across Great Barrier Reef catchments–paddock to reef integrated monitoring, modelling and reporting program. Marine pollution bulletin, 65(4-9), 117-127.

Smol, J. P., Wolfe, A. P., Birks, H. J. B., Douglas, M. S., Jones, V. J., Korhola, A., ... & Weckström, J. (2005). Climate-driven regime shifts in the biological communities of arctic lakes. Proceedings of the National Academy of Sciences, 102(12), 4397-4402.

Sohi, S. P. (2012). Carbon storage with benefits. Science, 338(6110), 1034-1035.

Song, M., & Pei, H. (2018). The growth and lipid accumulation of Scenedesmus quadricauda during batch mixotrophic/heterotrophic cultivation using xylose as a carbon source. Bioresource Technology, 263, 525-531.

Suganya, T., Varman, M., Masjuki, H. H., & Renganathan, S. (2016). Macroalgae and microalgae as a potential source for commercial applications along with biofuels production: A biorefinery approach. Renewable and Sustainable Energy Reviews, 55, 909-941.

Suresh, B., & Ravishankar, G. A. (2004). Phytoremediation—a novel and promising approach for environmental clean-up. Critical reviews in biotechnology, 24(2-3), 97-124.

Thomas, S. C., & Gale, N. (2015). Biochar and forest restoration: a review and meta-analysis of tree growth responses. New Forests, 46(5-6), 931-946.

Trainor, F. R. (1984). Indicator algal assays: laboratory and field approaches. Algae as Ecological Indicators, Academic Press, New York 1984. p 3-14, 6 tab, 22 ref.

Yamamoto, G., Omori, M., Hashida, T., & Kimura, H. (2008). A novel structure for carbon nanotube reinforced alumina composites with improved mechanical properties. Nanotechnology, 19(31), 315708.

Zheng, Y., Li, Z., Tao, M., Li, J., & Hu, Z. (2017). Effects of selenite on green microalga Haematococcus pluvialis: Bioaccumulation of selenium and enhancement of astaxanthin production. Aquatic Toxicology, 183, 21-27.

Zhou, X., Xia, L., Ge, H., Zhang, D., & Hu, C. (2013). Feasibility of biodiesel production by microalgae Chlorella sp.(FACHB-1748) under outdoor conditions. Bioresource technology, 138, 131-135.

Zhu, X., Chen, B., Zhu, L., & Xing, B. (2017). Effects and mechanisms of biochar-microbe interactions in soil improvement and pollution remediation: a review. Environmental pollution, 227, 98-115.

11

Recent Trend in Fish Farming An Overview

Samim Dullah[1]*, Rahul Gogoi[2] and Madhurjya Ranjan Sharma[2]

[1] *Department of Zoology Nanda Nath College, Titabar-78563, Assam, India*

[2] *Department of Agricultural Biotechnology, Assam Agricultural University Jorhat-785013, Assam, India*

Abstract

Fish is an important dietary protein consumed by the vast majority of the world's population. Fish are an important source of omega-3 fatty acids and several other bioactive molecules. Globally, there is increasing awareness among people about the health and nutritional benefits of a fish-based diet. Fish farming is considered a promising activity to meet nutritional requirements and provide health benefits to people worldwide. With the rising population, people are becoming heavily dependent on fish and fish-based products. This has led to the development of new tools and technologies for diversifying and intensifying fish farming. However, fish overproduction can affect aquatic ecosystems. Therefore, researchers are trying hard to develop better technologies, and the government is implementing policies related to fish farming to sustainably obtain higher fish production. This chapter aims to highlight the different fish farming techniques as well as biotechnology and technological innovations to obtain highly sustainable fish production. It also outlines future trends and perspectives for fish farming development.

Keywords: *omega-3 fatty acid, bioactive molecules, health, nutritional benefits, tools, technologies, policies, innovations*

1. Introduction

The current method of agricultural production is fish farming. A type of aquaculture known as "fish production" or "fish farming" entails the process of growing and producing a range of fish species. Proteins, vitamins (especially B2 and D), calcium, and other minerals are abundant in fish (Khalili, Tilami, and Sampels, 2018; Bostock *et al.*, 2010). In the past, fish output was limited to freshwater fish capture. Fish farming, also known as pisciculture, is becoming a more common and commonly utilized technique for fish production. The aquaculture and fish farming industry is one of the fastest-growing food production sectors globally. The production rate of farmed fish has increased from less than one million metric tons, as reported in the 1950s to about 110.1 million metric tons in 2016, which is valued at over $243 billion. In the recent years, production of farmed finfish has reached 57.5 million tonnes (that is equivalent to 146.1 billion USD), including 49.1 million tonnes (109.8 billion USD) from inland aquaculture and 8.3 million tonnes (36.2 billion USD) from marine culture in the sea and coastal aquaculture on the shore (Rito and Viegas, 2020). Nowadays, the majority of the world's population is reliant on the consumption of fish from aquaculture. Over the past few decades, finfish production through aquaculture and fish farming has increased remarkably, including both production and economic yield. Consequently, fish farming is becoming a major source of agricultural food in the present scenario.

Over time, considerable progress has been made in the farming of aquatic species, notably finfish. Now-a-days it has come to notice that there has been a necessity for a productive and regulated environment for the production of fish and also shellfish. There are different fish farming systems for raising finfish. The systems vary greatly in terms of facilities, interaction with other agricultural endeavors, and cultural techniques. Landponds are the most popular infrastructure for freshwater fish farming. However, in recent years, there has been rapid and considerable improvement in freshwater fish farming systems that are linked to agricultural systems. This has not only led to an increase in productivity and boosted resource efficiency, but also had an influence on the environment. Several innovative startups have been idealized, concentrating on creating innovations that can help obtain greater production of fish in a manner that causes the least negative effects on the environment. There are reports of businesses and startups utilizing RAS, AI, and IoT to advance land-based fish farming systems (Anani *et al.*, 2022). The use of such innovative techniques has helped to produce fish populations with a high density in a regulated manner. Advancements in RAS are also important in fish farming, including machine-learning techniques for the analysis of massive datasets produced by RAS operations, foam fractionation, bio-filtration, and UV sterilization technologies. The current goal of RAS in fish farming is to find long-term solutions to save fish populations and ecosystems. Desired results can be obtained using modern techniques based on scientific, ecological, technological, and economic principles. Determining good management practices (GMPs) that establish adequate procedures regarding the ideal stocking density, the intensity of cultivation, and the maximum amount to be used for both feed and chemical and organic fertilizers, among other things, is necessary for the sustainability and competitiveness of fish farming (Sebastian,

2009). Thus, it can operate with zero effluent and harvesting techniques that minimize the contribution of both solids and nutrients to the environment. This requires that the ponds be constructed correctly, and those that already exist are adjusted, allowing fish farming to grow in an ethical and competitive manner.

The responsible management of fish farming can potentially alleviate some of the strain on capture fisheries and generate income and jobs for coastal communities. However, as fish farming output increases, concerns about its effects on the ecosystem and wild fish species have also increased. The likelihood that the industry may encounter new biological, economic, and social issues increases as production increases. This could impact the sector's capacity to continue producing fish in a way that is morally and environmentally sound while also being productive. Therefore, it is critical that the industry strive to keep an eye on and manage these issues to prevent up-scaling possible issues when increasing output. This chapter discusses the different fish farming techniques and technological obstacles observed in fish farming. This highlights the sustainability and technological evolution of fish farming in China. It also discusses government steps and policies that can help fish farming in a sustainable manner.

2. Fish Farming-Culture methods

2.1 Cage System

Innovation occurs simultaneously with science and practice. The conventional approach to fish farming is constrained by the availability of water and area. There has been an increase in the number of people engaging in fish farming for economic benefits on a daily basis. However, the availability of ponds and reservoirs that are suitable for fish farming is limited. Building cages is the only way to overcome this predicament. In the field of fish farming, cage fish farming is a novel biotechnology that promotes the integrated use of water. Large reservoirs, ponds, and even small rivers with suitable hydrochemical environments are suitable places where beds are formed (Sangirova *et al.*, 2020). Cage farms can be set up in these locations, which can help in the production of fish in the market. Cage farms range in size, from family farms to commercial facilities. Since the cost of fish is higher than that of fish raised in a conventional manner, intensive fish farming is not particularly developed in many regions of the world. Using this approach, approximately 1–100 tons of fin-fish can be obtained. However, using the conventional method, only approximately 2 tons of fish can be produced (Chu *et al.*, 2020; NR *et al.*, 2022).

2.2 Pond System

A form of fish farming called pond culture involves cultivating aquatic creatures in ponds, including fish, shrimp, prawns, and shellfish. It has a history spanning thousands of years and is among the most established and extensively used types of fish farming. Ponds with freshwater, brackish water, or saltwater can all be used for pond aquaculture. To use this technique, one must have a small fish tank or pond. As water-containing fish waste is utilized to fertilize agricultural land, it is one of the most advantageous fish farming approaches. Fish farming systems in ponds have two primary varieties: extensive and intensive. In extensive systems, fish are fed by ponds' natural food sources. They are frequently employed in small-scale production, and are normally low-input, low-output systems. In intensive systems, higher production levels are possible because fish receive more nourishment. Compared with extensive systems, they usually demand more management and are more complex (Akankali *et al.*, 2002; Akankali *et al.*, 2011).

2.3 Integrated Recycling System

The Integrated Recycling System is founded on the idea that there is no waste and is a well-recognized technique for raising "pure" fish. The Integrated Recycling System uses hydroponic beds situated next to sizable plastic tanks that are housed inside a greenhouse. The reason it is named "no waste" is that the fish feed waste is used to provide nutrients to the plants grown in the hydroponic beds while the tank water is gradually pumped to the hydroponic system. This technology is adapted to nearly every temperate region because it is housed in a greenhouse. The discharge of salted water is an essential consideration for preserving the electrolyte balance of the fish. This technique uses sizable plastic fish tanks installed inside greenhouses. In addition, the hydroponic bed was situated close to the tank. People grow a variety of herbs, including parsley and basil, using the water from fish tanks (Oribhabor and Ansa 2006; Thi Da *et al.*, 2020).

2.4 Classic Fry Farming

Another name for this is a "Flow through system." Raised from eggs to fingerlings or fry, trout, and other sports fish are frequently transported to streams and released. Typically, fresh stream water is used to raise fry in long shallow concrete tanks. Commercial fish food in pellet form is fed to fry (Tan, 2016). Fish feed farming has always been a contentious topic. Numerous farmed fish species, such as carp, catfish, and tilapia, do not need to eat meat or fish products in their diet. Most salmon species, which are classified as top-level carnivores, are fed fish feed, some of which are typically obtained from wild fish (anchovies, menhaden, etc.). Carnivorous fish have been successfully fed vegetable-derived proteins instead of fishmeal; however, the diets have not been successfully supplemented with vegetable-derived oils. An additional concern pertains to the confinement of farmed fish in areas that are never observed in the wild, such as 50,000 fish in a 2-acre (8,100 m2) space, with each fish occupying less space than a typical bathtub. This can result in several types of contamination. Fish that are packed closely rub against one another and the sides of their cages, causing damage to their fins and tails, as well as various illnesses and infections, as well as stress (Marzban *et al.*, 2021).

3. Sustainability in Fish Farming

Increasing the stocking density of growing facilities is required to enhance the production of aquatic organisms. Currently, aquaculture supplies more than half of the world's seafood and has the fastest growth rate in the world. Fish farming is a promising method for producing proteins with a high nutritional value. Global aquaculture has expanded dramatically in recent decades, and has greatly improved food security and reduced poverty worldwide. This growth has been attributed to the use of new aquaculture technologies. In terms of facilities, interactions with other agricultural efforts, culture techniques, and aquaculture systems are highly variable. For freshwater aquaculture, land ponds are the most popular type of facility. However, in recent times, there have been notable and rapid improvements to freshwater aquaculture farming systems that are connected with agricultural systems. These improvements have had an effect on the environment, in addition to increasing productivity and maximizing resource use (FAO. The State of World Fisheries and Aquaculture 2020, Sustainability in Action; FAO, Rome, Italy 2020). Increasing the stocking density of growing facilities is required to enhance the production of aquatic organisms. High-density cultivation can have negative impacts on ecosystems, including disease outbreaks, unsustainable feeding, competition, and chemical and biological contamination.

3.1 Classical Methods

The use of reasonable densities, feeding strategies, and locations is among the most sustainable techniques for fish farming. This may be confirmed by examining fish cortisol levels, which is a sign of stress in farmed animals (Hanke *et al.*, 2020). When barramudi L. calcarifer is cultivated in a three-phase system in India, including larviculture, pre-growing, and final grow-out in cages, favorable technical indicators (fish survival, feed conversion, growth rate, and productivity), economic parameters (cost-benefit ratio, payback period, and internal rate of return), and indicators of livelihood security demonstrate that this system is technically and economically viable and socially acceptable (Kumaran *et al.*, 2021).

3.2 IMTA (Integrated Multi-Trophic Aquaculture)

The release of aquaculture effluents, which are primarily derived from foods used in cultivation, is another topic worth discussing. If improperly handled, these effluents can negatively impact the environment. This idea served as the foundation for the development of the integrated multi-trophic aquaculture (IMTA) concept, which uses extractive organisms such as seaweed and mollusks to absorb nutrients and particles from the surrounding environment while applying a simplified food web structure to a farming system of fed species such as prawns and fish (Carballeira Brana *et al.*, 2021). In this regard, fish farmers need to consider a number of aspects when developing an effluent treatment unit, such as a recirculation system (RAS) and the utilization of macrophytes or adsorbents. The goal is to reduce environmental contamination using suitable remediation techniques. In eutrophic conditions, there will be a greater input of aquatic ecosystem components, mostly nitrogen and phosphorus, which will result in an increase in primary production in water bodies (Ahmed *et al.*, 2022)

However, with alternate options such as aquaponics, other aquatic species can become sustainable. Currently, RAS knowledge and technology makes these systems inexpensive and viable only for the production of high-value species. Numerous current and upcoming developments in the generation of renewable energy will reduce the operational expenses of RAS. Aquaculture producers, biologists, and engineers need to work together to design and continuously improve every aspect of the RAS to reduce RAS costs. More knowledge regarding RAS technology can be obtained through study and field testing, along with a deeper understanding of the interactions between its various components. Using a prototype cultivation tank created by the company Lusalgae Ltd., situated in Figueira da Foz, Coimbra, Portugal, the wastewater from the cultivation tanks of the most diverse and nutrient-rich marine species can be used, for example, for the cultivation of marine macroalgae. This illustrates the cultivation of the red macroalga Calliblepharis jubata in an external environment. Because of the macroalgae's increased biomass and ability to absorb excess nutrients from fish farming, it also improves environmental quality and can be utilized to extract carrageenan, a phycocolloid that is crucial to the global food chain (Gomez-Zavaglia *et al.*, 2019).

4. Technological Evolution in Fish Farming

Fish farming has significantly improved because of technology that has made it possible to precisely monitor fish health, optimize feeding procedures, and reduce waste. Aquaculture systems linked to online servers and/or workstations with the most suitable software to manage and control the system are associated with industry expertise in engineering and computer science as well as multisensory schemes.

4.1 Aquaculture 4.0

Industry 4.0 is associated with engineering and computer science knowledge, coupled with multisensory schemes for aquaculture systems associated with online servers and/or workstations with the most appropriate software to manage and control the system, thereby contributing to improved aquaculture productivity and efficiency, while lowering the overall costs. Long-term solutions for increasing production (quantity and quality) while lowering costs and pollution in aquaculture can be found in aquaculture 4.0 technologies. Abiotic and biotic variables affect the aquaculture system, which has a significant impact on aquaculture production, because aquaculture can occur offshore or onshore. The development of 4.0 technologies and procedures is necessary to address the environmental demands posed by aquaculture sites and farmed species (Behroozi and Couturier, 2019). The ability to control and observe RTD on a cloud-based platform is one of the main advantages of aquaculture 4.0, especially for marine farms where cages cannot always be accessed promptly and when needed. The cloud-based system of onshore aquaculture features, which are available from any location, is highly praised by onshore fish farmers (Aquaculture 4.0: Applying industry strategy to fisheries management).

4.2 Recirculation Aquaculture Systems (RAS)

Recirculating aquaculture systems (RASs) are high-volume fish farming techniques that use less water and land. The design of an RAS should ensure that the key factors influencing fish productivity and water quality are properly balanced. Warm and cold water species require a variety of general water quality criteria, such as temperature, carbon dioxide, oxygen, total suspended particles, total ammonia, unionized ammonia, nitrite, and nitrate. Unlike other aquaculture production techniques, it is an intense, high-density fish culture. With this technology, fish are usually raised in indoor/outdoor tanks in a controlled environment, as opposed to the conventional practice of growing fish in open ponds and raceways. Water was filtered and cleaned using recirculating systems and recycled back into fish culture tanks. All species raised in aquaculture can be processed using this technology, which is based on the use of mechanical and biological filters. Only sufficient fresh water was added to the tanks to compensate for evaporation, splash out, and waste material flushing. After the system was filled with reconditioned water, no more than 10% of the system's total water volume was added (Badiola *et al.*, 2012).

4.3 Smart aquacultures (offshore and onshore)

One of the most recent trends in the sustainable development of aquaculture is smart aquaculture, which emphasizes automation and intelligence. Aquaculture has greatly benefited from the use of modern intelligent technology, which has reduced labor costs, increased productivity, and improved environmental friendliness. Since offshore aquaculture is still a relatively new industry, it requires the integration of new technologies, such as augmented reality and artificial intelligence, to enhance and automate a variety of distant tasks, such as feeding, sampling, monitoring, and surveillance. Artificial intelligence (AI) has a subset called machine learning, which uses algorithm models that have been trained to identify and learn characteristics from the data it observes. Smart aquaculture is a smart mode of production. It can be automated and managed remotely using robotics, the IoT, big data, artificial intelligence, 5G, cloud computing, and robots. Conversely, intelligent aquaculture can be run by robots that can oversee all the facilities, machinery, and equipment needed to run the entire system and produce effective results (Kassem *et al.*, 2021). Artificial intelligence (AI) is being used in aquaculture for drones, disease prevention, fish seed screening, routine stock checks, shrimp farming, smartphone applications, fish processing, open sea fishing, shrimp supply chain block chain technology, and fish conservation. Smart aquaculture, which is based on artificial intelligence advancements, can optimize every stage of the process, from breeding and nursery to the growth stages of the cultured species. It can also handle other processes, such as preparing the cultured water resource, controlling water quality, preparing feed, feeding, classifying, grading, counting, and cleaning the cultured systems. The ultimate objective of creating smart aquaculture is to achieve high aquaculture production to meet global demand, while also safeguarding the environment.

Water quality is the primary factor that makes aquaculture successful and efficient. Numerous water quality indicators, including temperature, turbidity, carbon dioxide, pH, alkalinity, ammonia, nitrite, and nitrate, have been found to have a significant direct or indirect impact on the survival and growth of cultured species. The most reliable parameters were pH, dissolved oxygen, and temperature. Aquaculture is among the many industries in which IoT has been used in the past few years. The use of IoT in aquaculture has sparked a new trend in the field's sustainable development, allowing the use of intelligent equipment with real-time connected water monitoring capabilities to enhance working conditions for aquaculture farmers (Kassem *et al.*, 2021). Feeding is a primary concern in aquaculture. The conventional approach calls for farmers to disperse food around the pond or concentrate it in one area, based on the eating habits of the cultivated species. Thus, IoT must be implemented in the feeding system to automatically regulate the amount of feed as well as feed. This has numerous advantages, including lowering water quality contamination in aquaculture, managing leftovers, and saving labor (Hilborn *et al.*, 2020).

4.4 Genomic Technologies

Diseases are a serious and urgent threat to the ability of the aquaculture industry to grow sustainably, especially when their effects worsen as a result of changing weather patterns, such as rising water temperatures. Although many strategies have been developed to tackle infectious diseases, such as vaccines and biosecurity measures, their effectiveness varies depending on the disease and species and is subject to numerous variables. Effective methods for managing and stopping disease outbreaks in aquatic animal species are provided by genetics and genomics. Genetics and genomics have the potential to yield long-term remedies by creating disease-resistant strains. Traditional genetic improvement programs integrate pedigree data with phenotypic assessment. Phenotypic measurements include measuring characteristics such as disease incidence, severity, or pathogen load in the host population (Davies *et al.*, 2009). More significantly, the sequencing of the genomes of aquaculture species has advanced our knowledge of the genetic components of the host that affect immune response, resistance, and susceptibility to infectious illnesses. Genome sequencing also makes it possible to choose people based on the information found in their DNA. Genomic selection is the name of this strategy (Meuwissen *et al.*, 2001).

Appearance characteristics in fish, or outer body attributes that influence customer acceptability at the point of sale, have gained relevance in commercial fish farming, as cultural success is closely linked to the management of these traits. Skin tone and body shape are the two most significant physical attributes. Significant genetic variation was found within populations when the genetic basis of these traits was examined in a variety of fish, suggesting that genetic improvement may be possible. Research to identify the primary or secondary genes responsible for the aesthetic qualities of commercial fish is expanding. Genes governing body form and skin color have been successfully identified in model fish (Colihueque and Araneda, 2014). However, fish producers must change and align their goals for selective breeding with consumer expectations to meet this challenge. The identification of genes that underlie body form and skin color or quantitative trait loci (QTLs), where continuous variation of the different attributes that comprise these traits is often found, is an antiquated and older method that might be employed to achieve this goal.

5. Feed and Nutrition in Fish Farming

Rising fish feed prices and the environmental effects of over-harvesting forage fish for feed and fish oil have resulted in an increase in herbivorous fish (carp and tilapia) and omnivorous fish (barramundi), which require substantially less fishmeal to produce protein. Furthermore, antibiotics and pesticides used on farmed fish may influence other marine species and human health (Araujo *et al.*, 2022). These nutrients and contaminants descend to the ocean floor and may affect biodiversity. Meanwhile, research to discover alternatives to fishmeal feed or methods to make it more sustainable is ongoing. Thus, finding the finest fish feed formula also includes attempting to attain the lowest feed conversion ratio—the amount of feed supplied–in relation to the weight acquired by the fish.

As raw feed materials are becoming more expensive and less sustainable, there is a growing need to find more affordable and sustainable raw feed sources, which makes the manufacturing of feed based on new components crucial. Currently, vegetable-and insect-based feeds can be cost-effective substitutes for conventional feeds (vegetables that do not compete with one another for animal feed or human consumption). Therefore, the use of bacteria, yeast, microalgae, macroalgae, and insects can replace forage fish, especially in high-value species such as salmonids, which are essential for the viability of fed aquaculture. Furthermore, since feed accounts for nearly half of the variable production costs in fish farming, nutrition plays a critical role in the industry. The creation of new, balanced commercial diets that promote the best possible growth and health of fish has resulted in a significant evolution in fish nutrition in recent years. To address the growing demand for affordable, secure, and premium fish and seafood, the aquaculture industry is growing owing to the development of innovative species-specific dietary formulae (Naylor *et al.*, 2021).

Overfeeding is another issue in aquaculture, which wastes valuable feed. Additional repercussions include increasing bacterial loads, low dissolved oxygen levels, elevated biological oxygen demand, and water pollution. Generally speaking, fish should only be fed as much food as they can quickly consume, that is, within five–ten minutes. A good general rule of thumb is to feed the fish approximately 80% of their desired amount (satiation). In this strategy, you feed the fish as much as they can eat in one day on a regular basis, potentially twice a month. Consequently, new meals are more nutritionally beneficial (Dauda *et al.*, 2019).

This is due to antibiotic and pharmacological restrictions on fish cultivation. The industry's future development is strongly based on the use of natural resources. In aquaculture, nutraceuticals are used to enhance food conversion, growth performance, disease resistance, and product safety for human consumption. Probiotics increase overall vigor, improve health, encourage resilience to disease, lessen sensitivity to stress, and increase growth and feed conversion. Currently, land-based sources, rather than fish, provide the majority of nutraceuticals. However, it is anticipated that host-associated (autochthonous) nutraceuticals will be more resilient in fish gastrointestinal tracts and, hence, may offer longer-lasting advantages to the host. Nutraceutical candidates are commonly assessed in vitro; however, switching to in vivo testing is challenging (Bianchi *et al.*, 2022).

Although administering adequate doses of immunostimulants or immunomodulatory agents increases resistance to various diseases and improves animal health, some studies have shown that administering an optimal dose of immunostimulants is critical for achieving an effective response. These effects have been demonstrated in the production of Asian seabass (Lates calcarifer), and the addition of Kappaphycus alvarezii (Rhodophyta) demonstrated great immunostimulatory activity, making it a good immunostimulant/immunomodulatory agent in the fish aquaculture industry, although more experiments are required (Ali *et al.*, 2022).

Aquaculture appears to be a promising tool in the form of natural substances that can modulate the response of fish to stressors, when paired with well-managed cultivation. This approach prevents adverse effects and reduces fish mortality during production.

6. Biosecurity in Fish Farming-Government policies

As defined by the FAO, biosecurity is an all-encompassing and integrated approach that includes environmental issues and significant risks to human, animal, and plant life and health. They also include legislative and regulatory frameworks. A wide range of topics have been covered, including invasive alien species, zoonoses, food safety, and the introduction and treatment of diseases in plants and animals. Along with their byproducts, it also discusses the introduction and spread of modified living organisms. Ignoring biosecurity can lead to disease outbreaks, which can hinder farm access to markets, negatively impact farm output, and pose health risks to humans. Additionally, they can give fish a bad taste and appearance. Farmers' cash returns are naturally reduced. Globalization, for instance, has raised the risk of disease spread because aquatic animal trafficking has increased in volume, diversity, and socioeconomic importance. Building a successful biosecurity plan involves several steps. These processes include prioritizing and identifying hazards; assessing risks and their implications; identifying, mitigating, managing, and correcting critical control points that allow diseases to enter or exit the epidemiological unit; creating a backup plan in case a disease is found there through disease surveillance, monitoring, and determining the status or freedom of the disease within the epidemiological unit; and routine review procedures. These protocols are tested by veterinarians, and the government veterinary authorities should assess and certify them (Scarfe and Palic, 2020).

It has been shown that aquaculture has the ability to support the production of food on a worldwide scale that benefits society. However, if the aquaculture blue revolution is pushed to solve the growing global food security challenges, remedial measures to lessen its adverse effects also need to be created. Aquaculture has increased the amount of animal protein consumed per person and increased global food production; however, in certain cases, it has also depleted resources that support regional and global food security. Certain aquaculture technologies are considered hazardous to food security (Stentiford *et al.*, 2020).

Individual farms are the focus of most aquaculture laws and certification schemes. Many producers in one area may have cumulative effects on the ecosystem, even if everyone abides by regulations. Examples of these effects include fish illness and water pollution. Zoning and spatial planning can ensure that aquaculture operations remain within the environmental carrying capacity of the surrounding area, and can also help minimize conflicts over resource use. To reduce the risk of disease and assist in controlling the effects on the ecosystem, Norway, for example, has zoning restrictions that ensure that salmon farmers are not overly concentrated in one area (Lester *et al.*, 2018).

Many rice fields have been turned into fishponds in China, Thailand, and Vietnam, and China has since outlawed this practice due to concerns about national food security. More importantly, it signifies a shift from growing a primary food crop for the community to growing a commodity in the export market. As a result, some small farmers who have turned productive rice fields into fish farms are concerned about food security.

Beveridge *et al.*, (2013) stated that the importance of aquaculture to food security depends on a number of factors, including fish size, product price, culture species, and availability for low-income consumers. Numerous commercial and government initiatives can provide farmers with incentives to practice sustainable aquaculture. For example, prawn farmers operating lawfully in aquaculture zones in Thailand have access to free wastewater treatment, water supply, and training from the Thai government. Small-scale farmers have benefited from low-interest loans and tax advantages provided by the government, which has helped them embrace advanced technology that has increased productivity and reduced the need for new land clearance. As a result, to encourage farmworkers to use biosecurity measures, trustworthy and respectable sources such as veterinarians should educate them on the subject so that they are aware of both the advantages and disadvantages of doing so.

Regulations, production, external exposure, and internal infrastructure changes need to be considered when evaluating and updating a biosecurity policy. Building a written biosecurity plan based on risk assessment and using audits to gauge the plan's effectiveness in mitigating risks and hazards is the most efficient way to develop strong biosecurity at aquaculture facilities. When evaluating the many components and actions in the plan, both separately and in conjunction with other biosecurity farm programs, the grading scheme is crucial in ascertaining the relative importance of each. Unless farm workers are properly trained and accept biosecurity as a standard operating procedure, no biosecurity plan will be successful. Farmers must work together openly on a regional, national, and global scale for biosecurity to be financially viable. It is essential to disclose vital facts and information about the state of local health, especially the incidence of rising death rates and the frequency of infectious diseases. The industry cannot successfully prevent and control disease outbreaks unless stakeholders collaborate transparently with one another.

Conclusion and Future Perspectives

Even though the world's fish production is trending upward, it is unable to meet the demands of a global population that is growing at an unabated rate. Fisheries and aquaculture have significantly contributed to the GDP of both countries and the world. In addition to eliminating post-harvest losses, this will ensure food and nutritional security, create jobs, raise revenue, and provide hygienic fish. Prospects for fish farming in the future can also be considered as a means of providing millions of people worldwide. In terms of production and marketing, the overall fish productivity fell short of expectations. It is important to support market infrastructure, storage facilities, insulated cars, pricing regulations, the provision of incentives, training, extension initiatives, and an off-site fish market. To ensure that fisheries thrive financially, a variety of agricultural organizations, aquacultural agencies, trade specialists, policy-makers, and producers should develop plans and strategies. Additionally, current fish growing techniques, such as polyculture and Paddycum fish culture, should be promoted to farmers. For producers not to experience financial losses following production, marketing networks need to be handled well.

References

Ahmad, A.L.; Chin, J.Y. and Mohd Harun, M.H.Z.; Low, S.C. Environmental impacts and imperative technologies towards sustainable treatment of aquaculture wastewater: A review. J. Water Process Eng. 2022, 46, 102553 https://doi.org/10.1016/j.jwpe.2021.102553

Akankali, J. A., Abowei, J. F. N., and Eli, A. (2011). Pond fish culture practices in Nigeria. Advance Journal of Food Science and Technology, 3(3), 181-195.

Akbar Ali, I., Radhakrishnan, D.K. and Kumar, S. (2022). Immunostimulants and Their Uses in Aquaculture. In: Balasubramanian, B., Liu, WC., Sattanathan, G. (eds) Aquaculture Science and Engineering. Springer, Singapore. https://doi.org/10.1007/978-981-19-0817-0_11

Akeem Babatunde Dauda, Abdullateef Ajadi, Adenike Susan Tola-Fabunmi and Ayoola Olusegun Akinwole. Waste production in aquaculture: Sources, components and managements in different culture systems, Aquaculture and Fisheries, Volume 4, Issue 3, 2019, Pages 81-88, ISSN 2468-550X, https://doi.org/10.1016/j.aaf.2018.10.002

Anani, O. A., Adetunji, C. O., Olugbemi, O. T., Hefft, D. I., Wilson, N., and Olayinka, A. S. (2022). IoT-based monitoring system for freshwater fish farming: Analysis and design. In AI, Edge and IoT-based Smart Agriculture (pp. 505-515). Academic Press.

Araujo, G.S.; Silva, J.W.A.d.; Cotas, J. and Pereira, L. Fish Farming Techniques: Current Situation and Trends. J. Mar. Sci. Eng. 2022, 10, 1598. https://doi.org/10.3390/jmse10111598

Behroozi, L. and Couturier, M.F. Prediction of water velocities in circular aquaculture tanks using an axisymmetric CFD model. Aquac. Eng. 2019, 85, 114–128 https://doi.org/10.1016/j.aquaeng.2019.03.005

Beveridge MC, Thilsted SH, Phillips MJ, Metian M, Troell M and Hall SJ. Meeting the food and nutrition needs of the poor: the role of fish and the opportunities and challenges emerging from the rise of aquaculture. J Fish Biol. 2013 Oct;83(4):1067-84. doi: 10.1111/jfb.12187. Epub 2013 Aug 30. PMID: 24090563

Bianchi, M., Hallström, E. and Parker, R.W.R. (2022). Assessing seafood nutritional diversity together with climate impacts informs more comprehensive dietary advice. Commun Earth Environ 3, 188. https://doi.org/10.1038/s43247-022-00516-4

Bostock, J., McAndrew, B., Richards, R., Jauncey, K., Telfer, T., Lorenzen, K., and Corner, R. (2010). Aquaculture: global status and trends. Philosophical Transactions of the Royal Society B: Biological Sciences, 365(1554), 2897-2912.

Carballeira Braña, C.B.; Cerbule, K.; Senff, P. and Stolz, I.K. Towards Environmental Sustainability in Marine Finfish Aquaculture. Front. Mar. Sci. 2021, 8, 666662. https://doi.org/10.3389/fmars.2021.666662

Chu, Y. I., Wang, C. M., Park, J. C., and Lader, P. F. (2020). Review of cage and containment tank designs for offshore fish farming. Aquaculture, 519, 734928.

Colihueque, N. and Araneda, C. 2014. Appearance traits in fish farming: Progress from classical genetics to genomics, providing insight into current and potential genetic improvement. Front. Genet. 5, 251. https://doi.org/10.3389/fgene.2014.00251

David Scarfe and Dusan Palic. Aquaculture biosecurity: Practical approach to prevent, control, and eradicate diseases. Aquaculture Health Management. Academic Press, 2020, Pages 75-116, ISBN 9780128133590, https://doi.org/10.1016/B978-0-12-813359-0.00003-8

Davies, G.; Genini, S.; Bishop, S.C. and Giuffra, E. 2009. An assessment of opportunities to dissect host genetic variation in resistance to infectious diseases in livestock. Animal 3, 415–436. 10.1017/S1751731108003522

FAO, 2020. The State of World Fisheries and Aquaculture 2020 Sustainability in Action; FAO: Rome, Italy.

Gomez-Zavaglia A, Prieto Lage MA, Jimenez-Lopez C, Mejuto JC and Simal-Gandara J. 2019. The Potential of Seaweeds as a Source of Functional Ingredients of Prebiotic and Antioxidant Value. Antioxidants (Basel). Sep 17;8(9):406. doi: 10.3390/antiox8090406. PMID: 31533320; PMCID: PMC6770939.

Hanke, I.; Hassenrück, C.; Ampe, B.; Kunzmann, A.; Gärdes, A. and Aerts, J. 2020. Chronic stress under commercial aquaculture conditions: Scale cortisol to identify and quantify potential stressors in milkfish (Chanos chanos) mariculture. Aquaculture 526, 735352.

Hilborn, R.; Amoroso, R.O.; Anderson, C.M.; Baum, J.K.; Branch, T.A.; Costello, C.; De Moor, C.L.; Faraj, A.; Hively, D. and Jensen, O.P.; 2020. Effective fisheries management instrumental in improving fish stock status. Proc. Natl. Acad. Sci. USA, 117, 2218–2224. https://doi.org/10.1073/pnas.1909726116

Kassem, T.; Shahrour, I.; El Khattabi, J. and Raslan, A. 2021. Smart and Sustainable Aquaculture Farms. Sustainability, 13, 685. https://doi.org/10.3390/su131910685

Khalili Tilami, S., and Sampels, S. (2018). Nutritional value of fish: lipids, proteins, vitamins, and minerals. Reviews in Fisheries Science & Aquaculture, 26(2), 243-253.

Kumaran, M.; Vasagam, K.P.K.; Kailasam, M.; Subburaj, R.; Anand, P.R.; Ravisankar, T.; Sendhilkumar, R.; Santhanakumar, J. and Vijayan, K.K. 2021. Three-tier cage aquaculture of Asian Seabass (Lates calcarifer) fish in the coastal brackishwaters—A techno-economic appraisal. Aquaculture, 543, 737025.

Lester, S.E., Stevens, J.M. and Gentry, R.R. (2018). Marine spatial planning makes room for offshore aquaculture in crowded coastal waters. Nat Commun 9, 945 https://doi.org/10.1038/s41467-018-03249-1

Maddi Badiola, Diego Mendiola and John Bostock, 2012. Recirculating Aquaculture Systems (RAS) analysis: Main issues on management and future challenges, Aquacultural Engineering, Volume 51, Pages 26-35, ISSN 0144-8609, https://doi.org/10.1016/j.aquaeng.2012.07.004

Marzban, A., Elhami, B., and Bougari, E. (2021). Integration of life cycle assessment (LCA) and modeling methods in investigating the yield and environmental emissions final score (EEFS) of carp fish (Cyprinus carpio) farms. Environmental Science and Pollution Research, 28(15), 19234-19246.

Meuwissen, T.H.; Hayes, B.J. and Goddard, M.E. 2001. Prediction of total genetic value using genome-wide dense marker maps. Genetics, 157, 1819–1829. 10.1093/genetics/157.4.1819

Mmochi, A. J., Dubi, A. M., Mamboya, F. A., & Mwandya, A. W. (2002). Effects of Fish Culture on Water Quality of an Integrated Mariculture Pond System.

Naylor, R.L., Hardy, R.W. and Buschmann, A.H. (2021). A 20-year retrospective review of global aquaculture. Nature 591, 551–563. https://doi.org/10.1038/s41586-021-03308-6

NR, M., XG, A., SU, I., and JJ, S. (2022). Biotechnological aspects of cage growing of African catfish in the conditions of Uzbekistan. Deutsche Internationale Zeitschrift für Zeitgenössische Wissenschaft, (41).

Oribhabor, B. J., and Ansa, E. J. (2006). Organic waste reclamation, recycling and re-use in integrated fish farming in the Niger Delta. Journal of Applied Sciences and Environmental Management, 10(3), 47-53.

Rito, J., and Viegas, I. (2020). Fish farming. Life Below Water, 1-9.

Sangirova, U., Khafizova, Z., Yunusov, I., Rakhmankulova, B., and Kholiyorov, U. (2020). The benefits of development cage fish farming. In E3S Web of Conferences (Vol. 217, p. 09006). EDP Sciences.

Sebastian, A. (2009). Development of safety and quality management system in shrimp farming (Doctoral dissertation, Cochin University of Science and Technology).

Stentiford, G.D., Bateman, I.J. and Hinchliffe, S.J. (2020). Sustainable aquaculture through the One Health lens. Nat Food 1, 468–474. https://doi.org/10.1038/s43016-020-0127-5

Tan, K. L. (2016). 3D Fish Culture and Monitoring System. IRC.

Thi Da, C., Anh Tu, P., Livsey, J., Tang, V. T., Berg, H., and Manzoni, S. (2020). Improving productivity in integrated fish-vegetable farming systems with recycled fish pond sediments. Agronomy, 10(7), 1025.

12

Zero Tillage Fish Farming and its Importance

Avik Bhanja[1], Prabir Sahoo[2*], Sagar Samanta[3], Basudev Mandal[4] and Pijush Payra[5]

[1]Department of Fishery Sciences, Vidyasagar University, Midnapore-721102 West Bengal, India

[2]Department of Coastal Aquaculture, Centre of Advanced Study in Marine Sciences Annamalai University, Cuddalore-608502, Tamil Nadu, India

[3]Department of Fisheries Science, Medinipur City College, West Medinipur-721129 West Bengal, India

[4]Narajole Raj College, Narajole, Paschim Medinipur-721211, West Bengal, India

[5]Department of Industrial Fish and Fisheries (HOD), ACM and Fisheries Science (PG), Ramnagar College, Depal-721453, West Bengal, India

Abstract

Zero tillage fish farming has emerged as a transformative and sustainable approach to aquaculture, revolutionizing traditional methods by minimizing soil disturbance and preserving the natural pond ecology. This chapter explores the principles, benefits, and importance of zero-tillage fish farming, emphasizing its role in environmental sustainability, food security, and mitigating the impact of climate change. Zero tillage enhances productivity and efficiency in fish cultivation by maintaining optimal water quality and supporting biodiversity. This method is a beacon of responsible resource management, offering a promising solution to the challenges of a growing global population. As the world seeks more sustainable practices, zero tillage fish farming presents itself as a crucial tool in reshaping the future of aquaculture towards a resilient and ecologically conscious industry.

Keywords: *Zero tillage, Fish farming, Importance, Sustainability*

1. Introduction

The increasing global population has intensified the challenges associated with land and freshwater scarcity (Dutta, 2019). The unpredictable supply of water bodies has prompted a significant shift among farmers towards the utilization of private tube wells, exerting increased pressure on groundwater resources (Ahmad *et al.*, 2003; Ahmad *et al.*, 2007). Due to inadequate water management practices and excessive extraction of groundwater, leading to a decline in water tables, irrigation is increasingly exerting a detrimental influence on the ecosystem (Pingali and Shah 2001; Qureshi *et al.*, 2003). Therefore, agricultural technologies that contribute to water conservation, cost reduction, and increased productivity are gaining increasing importance (Gupta *et al.*, 1997; Gupta *et al.*, 2002).

Zero tillage fish farming, a groundbreaking approach in aquaculture, has emerged as a sustainable and efficient method for cultivating fish without disturbing the aquatic environment. Zero tillage fish farming is an innovative and sustainable aquaculture method that minimizes soil disturbance and preserves natural pond ecology (Hussan *et al.*, 2019). In contrast to traditional practices involving pond bottom dredging, zero tillage maintains water quality, supports biodiversity, and enhances overall environmental sustainability (FAO AQUASTAT 2009). By avoiding disruptive practices, this approach promotes self-sustaining aquatic ecosystems, contributing to increased productivity, cost-effectiveness, and resilience in fish cultivation. Zero tillage fish farming represents a significant shift towards responsible and eco-friendly aquaculture, addressing challenges such as environmental degradation, food security, and the impact of climate change. This chapter explores the principles, benefits, and importance of zero-tillage fish farming, shedding light on its potential to transform the aquaculture industry.

2. Concept of Zero Tillage Farming

Edward H. Faulkner introduced the concept of modern no-till farming in the 1940s (Derpsch, 2011). In the late 1940s, the rise in minimal tillage became popular and coincided with the development of plant growth regulators during World War II (Phillips and Phillips, 1984). In the 1960s, the M.A. Sprague documented the use of chemicals for pasture rejuvenation, marking a shift away from traditional tillage practices (Phillips and Phillips, 1984).

No-till or zero-tillage farming is an agricultural methodology dedicated to the cultivation of crops without resorting to soil disturbance through tillage (Erenstein *et al.*, 2007; Eurostat, 2020; Cerdà *et al.*, 2020; Cerdà *et al.*, 2020). This technique offers various potential advantages such as enhancing water infiltration into the soil, promoting the retention of organic matter, and optimizing nutrient cycling processes (Govt. of Assam, 2018; El-Shater *et al.*, 2020; Wikipedia, 2023). The zero-tillage concept was first introduced nearly two decades ago and aimed to assist farmers in minimizing tillage expenditures and expediting the planting season for wheat and other Rabi crops (Game, 2021).

Zero-tillage fish farming, also known as no-till aquaculture, involves cultivating fish in a manner that minimizes soil disturbance and environmental impacts (Hussan *et al.*, 2019). Unlike traditional aquaculture methods, which often involve dredging or tilling the pond bottom, zero tillage emphasizes maintaining the natural balance of the aquatic ecosystem (Muruganandam, 2012a; Muruganandam, 2012b). Recirculatory aquaculture systems (RAS), Biofloc Culture, Aquaponics, Integrated Multitrophic Aquaculture (IMTA), Cage Culture, Pen Culture, Crop Rotation (Berg, 2002; Eer *et al.*, 2004; Derpsch, 2011; NFDB. 2016; Yavuzcan *et al.*, 2016; Vigyan Ashram, 2020; Balami, 2021; Yang *et al.*, 2023) are ideal examples of zero-tillage fish farming practices.

3. Different Process to Maintain Zero Tillage in Fish Farming

Maintaining zero tillage in fish farming involves implementing practices that minimize soil disturbance and preserve the natural balance of aquatic ecosystems (Derpsch *et al.*, 2010). The following are the key guidelines for maintaining zero tillage in fish farming (Figure 1).

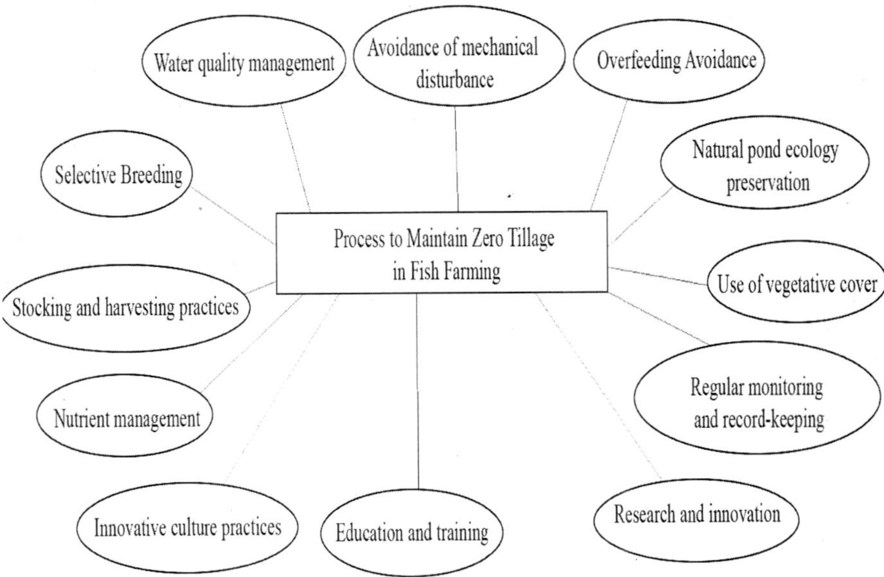

Fig. 1. Different Process to Maintain Zero Tillage in Fish Farming

3.1 Water quality management

It is essential to regularly monitor and maintain optimal water quality parameters, including oxygen levels, temperature, and pH, in no-tillage fish farms. The implementation of aeration systems is essential to ensure adequate oxygenation, especially in deep ponds (SAFDC, 1980).

3.2 Avoidance of mechanical disturbance

Avoidance of any mechanical disturbance is one of the major criteria for maintaining no-tillage practices in farming (Reicosky and Saxton, 1996). In fish farming, it is important to avoid dredging or tilling the pond bottom to prevent the disruption of natural sediments and adverse effects on water quality (Malone and Polyakov, 2019). Implementing a proper pond design that minimizes the need for mechanical intervention is crucial for maintaining the zero-tillage approach.

3.3 Overfeeding Avoidance

Overfeeding can lead to excess waste in water. It is important to feed fish in appropriate amounts to prevent food from decomposing and causing water quality issues (Craig and Kuhn 2017).

3.4 Selective breeding

Fish are selectively bred for traits such as disease resistance and fast growth to reduce the need for antibiotics and other chemical interventions (National Research Council, 1999; Shinde and Sukhdhane, 2023).

3.5 Natural pond ecology preservation

Ensuring an adequate presence of beneficial microorganisms, plankton, and other organisms promotes the development of the ecology of a water body (Hill *et al.*, 2021).

Avoiding the use of chemicals that can disturb the balance of the ecosystem (Bashir *et al.*, 2020) is essential for zero-tillage farming.

3.6 Stocking and harvesting practices

Appropriate stocking densities for pond size and carrying capacity must be applied to prevent overcrowding and stress on fish (Centurion University, 2020).

Selective harvesting methods should be adopted to avoid unnecessary disturbance to the pond bottom (Kungvankij *et al.*, 1986).

3.7 Use of vegetative cover

Emphasis should be placed on the growth of aquatic plants around the edges of ponds to provide natural cover for fish and to prevent bank erosion.

Floating plants can be used to provide shade and shelter to fish, contributing to the overall stability of the pond (Patnaik and Ramaprabhu, 1985; Vincent *et al.*, 2017).

3.8 Nutrient management

Monitoring nutrient levels in the water prevents excessive accumulation, which can lead to water quality issues (Game, 2021). Natural nutrient cycling practices, such as using organic matter or biological filters, maintain nutrient balance (Ujoh *et al.*, 2016).

3.9 Regular monitoring and record-keeping

It is important to conduct regular water quality tests and ecological assessments to track the changes in the pond environment (Eer *et al.*, 2004). No-till fish farming also requires detailed records of stocking densities, feeding regimes, and any interventions to evaluate the impact of no-till fish farming on the ecosystem.

3.10 Innovative culture practices

Different innovative culture practices such as aquaponics, hydroponics, Integrated Polyculture, Integrated Multitrophic Aquaculture (IMTA), and Recirculatory Aquaculture System (RAS) Biofloc Practices can eliminate the need for soil, thus avoiding soil tillage (Panigrahi *et al.*, 2019; Vigyan Ashram, 2020). These systems reduce environmental impacts and enhance sustainability. Such innovative cultural practices can boost the concept of zero tillage farming in fisheries.

3.11 Education and training

Educating and training agricultural workers about the principles and significance of zero-tillage systems, along with fostering a culture of environmental stewardship and awareness among individuals engaged in fishery activities (FAO, 2014; Prakash, 2024; Machuka, 2024), can establish the groundwork for a sustainable fisheries system with zero tillage practices.

3.12 Research and innovation

Research and innovation play vital roles in advancing and optimizing no-tillage fish farming practices, contributing to the sustainability and efficiency of aquaculture (Game 2021). Research and innovation help design and refine no-tillage fish farming systems, considering different factors such as pond structure, water circulation, integration with plant components, and development of efficient and sustainable technologies.

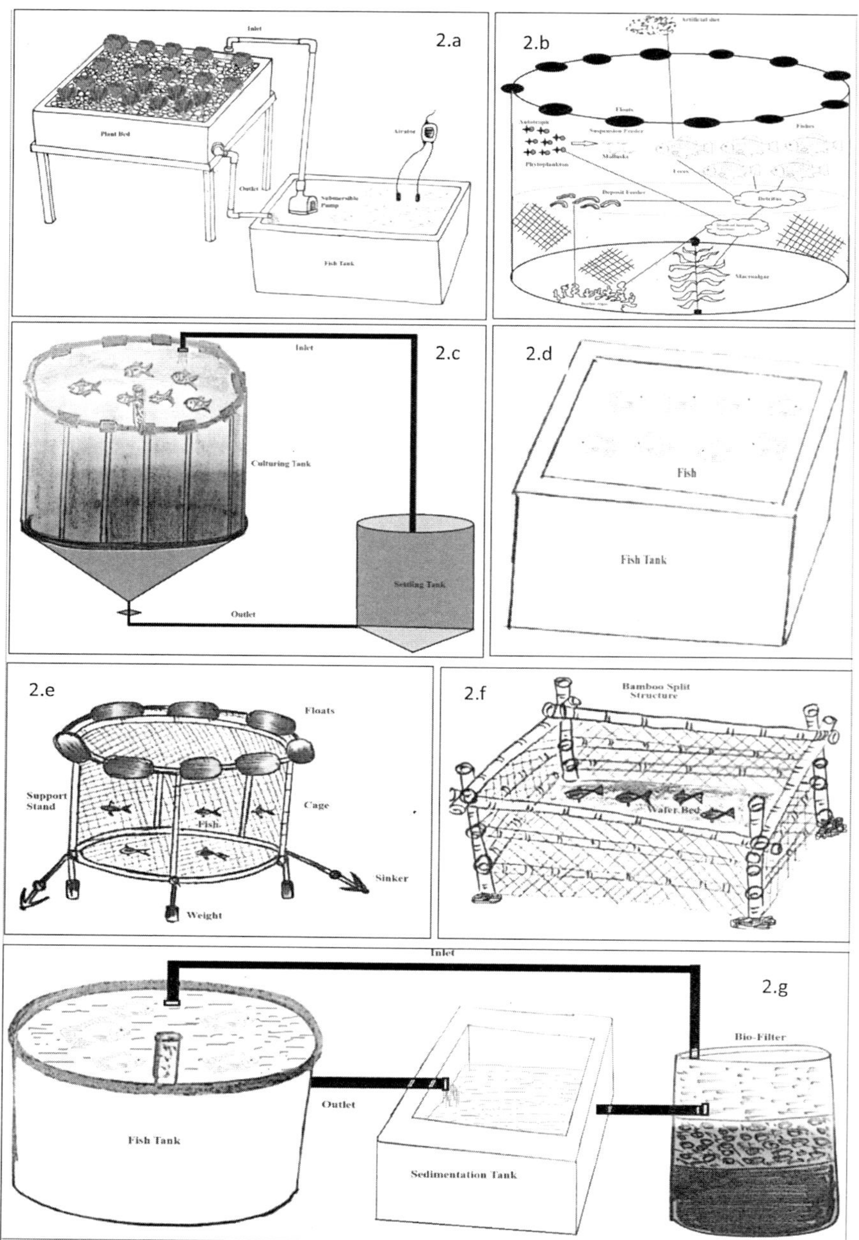

Fig. 2. Some fish farming practices related to the zero-tillage concept are shown in Fig. 2. a. Aquaponics farming; Fig 2. b. IMTA; Fig 2.c. Biofloc farming; Fig 2.d. Tank farming; Fig 2.e. Cage Culture; Fig 2.f. Pen Culture; Fig 2.g. RAS farming

4. Benefits of zero tillage fish farming

4.1 Improved water conservation

Zero tillage significantly reduces water usage by maintaining the integrity of pond ecosystems (Cherlinka, 2023). This method minimizes water loss through evaporation and seepage, thereby contributing to sustainable water conservation practices (Flower *et al.*, 2008).

4.2 Enhanced biodiversity

By preserving the natural pond ecology, zero tillage fosters a diverse and balanced ecosystem (Ujoh *et al.*, 2016). This not only benefits the target fish species, but also supports the growth of other aquatic organisms, creating a healthier and more resilient environment (Hussan *et al.*, 2019).

4.3 Increased productivity and efficiency

Zero tillage fish farming has been shown to enhance the productivity and efficiency of aquaculture. This method reduces the need for input costs (Tumma and Chandrika, 2023) in the dredging or tilling of culture ponds, supplemental feeds, and energy-intensive practices, resulting in a more cost-effective and environmentally friendly approach to fish cultivation.

5. Importance of Zero Tillage Fish Farming

5.1 Environmental sustainability

As global concerns about environmental degradation and resource depletion intensify, zero-tillage fish farming stands out as an environmentally sustainable alternative to traditional aquaculture methods. By preserving natural habitats and reducing the ecological footprint, this approach aligns with the principles of responsible and sustainable aquaculture (Savage 2017).

5.2 Food security and economic viability

As the world's population continues to grow, ensuring food security is becoming a pressing challenge (Duboc *et al.*, 2011). Zero-tillage fish farming contributes to meeting this challenge by providing a sustainable and efficient method for fish production. Additionally, its economic viability makes it an attractive option for small-scale commercial ventures (Baker *et al.*, 2005).

5.3 Mitigating climate change impact

Traditional aquaculture practices often contribute to greenhouse gas emissions and environmental degradation (Datta 2019). Zero tillage fish farming, with its focus on preserving natural ecosystems, helps mitigate the impact of climate change by reducing carbon emissions and promoting a more resilient aquatic environment (Duboc *et al.*, 2011; Encyclopedia Britannica, 2023).

Conclusion

Zero tillage fish farming represents a paradigm shift in aquaculture, emphasizing sustainability, efficiency, and environmental stewardship. As the importance of responsible resource management becomes increasingly evident, adopting zero tillage practices can pave the way for a more sustainable and resilient future in the aquaculture industry. Through continued research, innovation, and widespread adoption, zero-tillage fish farming has the potential to revolutionize how we cultivate and harvest fish, ensuring a healthier planet for generations to come.

References

Ahmad, M. D., Turral, H., Masih, I., Giordano, M. and Masood, Z. 2007. Water saving technologies: Myths and realities revealed in Pakistan's rice-wheat systems. IWMI Research Report 108. Colombo, Sri Lanka: International Water Management Institute.

Baker, C.J., Saxton, K.E., Ritchie, W.R., Chamen, W.C.T., Reicosky, D.C., Ribeiro, M.F.S., Justice S.E. and Hobbs, P.R. 2005. No-tillage Seeding in Conservation Agriculture (Second Edition). C.J. Baker and K.E. Saxton (Eds.). Food and Agriculture Organization of the United Nations. CAB International, Wallingford, UK.

Bashir, I., Lone, F.A., Bhat, R.A., Mir, S.A., Dar, Z.A. and Dar, S.A. 2020. Concerns and Threats of Contamination on Aquatic Ecosystems. Bioremediation and Biotechnology (Hakeem et al., eds.), pp.1-26. PMCID: PMC7121614.

Berg, H. 2002. Rice monoculture and integrated rice-fish farming in the Mekong Delta, Vietnam—Economic and ecological considerations. Ecological Economics, 41(1), pp. 95-107. https://doi.org/10.1016/S0921-8009(02)00027-7

Centurion University. 2020. Soil and water quality parameters in aquaculture. Unit 9. https://courseware.cutm.ac.in

Cerdà, A., Rodrigo-Comino, J., Yakupoglu, T., Dindaroglu, T., Terol, E., Mora-Navarro, G., Arabameri, A., Radziemska, M., Novara, A., Kavian, A., Vaverková, M.D., Abd-Elmabod, S.K., Hammad, H.M. and Daliakopoulos, I.N. 2020. Tillage Versus No-Tillage. Soil Properties and Hydrology in an Organic Persimmon Farm in Eastern Iberian Peninsula. Water, 12, 1539, p.16.

Cherlinka, V. 2023. No-Till Farming: Way to More Sustainable Agriculture. EOS Data Analytics. https://eos.com/blog/no-till-farming/

Craig, S. and Kuhn, D.D. 2017. Understanding Fish Nutrition, Feeds, and Feeding. Virginia Cooperative Extension (Publication 420-256). Virginia Tech, Virginia State University, Petersburg, P.16.

Datta, S. 2009. Soil and Water Interaction and Its Importance in Aquaculture. CIFE (Central Institute of Fisheries Education), Kolkata Centre, West Bengal, India. http://www.scribd.com/doc/21749742/Soil-Water-Interaction-and-Its-Implications-in-Aquaculture

Derpsch, R., Friedrich, T., Kassam, A. and Hongwen, L. 2010. Current status of adoption of no-till farming in the world and some of its main benefits. Int J Agric & Biol Eng, 3(1), pp.1-25.

Derpsch, R. 2011. Historical review of no-tillage cultivation of crops. No-Tillage, Sustainable Agriculture in the New Millennium. MAG- GTZ Soil Conservation Project. Retrieved on January 2nd, 2024. http://www.rolf-derpsch.com/en/no-till/historical-review/

Duboc, O., Zehetner, F. and Gerzabek, M. H. 2011. Recent Developments of No-Till and Organic Farming in India: Is a Combination of These Approaches Viable? Journal of Sustainable Agriculture, 35, pp.576–612.

Dutta, A. 2019. Economics of Water in India: Why does it Matter? Economics, 3(5).

Eer, A.V., Schie, T.V. and Hilbrands A. 2004. Small-scale freshwater fish farming. Agromisa Foundation, Wageningen, Netherland, p.79.

El-Shater, T., Mugera, A. and Yigezu, Y. A. 2020. Implications of Adoption of Zero Tillage (ZT) on Productive Efficiency and Production Risk of Wheat Production. Sustainability, 12, 3640.

Encyclopaedia Britannica. 2023. No-till agriculture. Encyclopaedia Britannica, Inc. Retrieved on January 2nd, 2024. https://www.britannica.com/topic/till-less-agriculture

Erenstein, O., Farooq, U., Malik R.K. and Sharif M. 2007. Adoption and impacts of zero tillage as a resource conserving technology in the irrigated plains of South Asia. Colombo, Sri Lanka: International Water Management Institue. p. 55 (Comprehensive Assessment of Water Management in Agriculture Research Report 19)

Eurostat. 2020. Agri-environmental indicator – tillage practices. https://ec.europa.eu/eurostat/statistics-explained/index.php?title=Agri-environmental_indicator_-_tillage_practices

FAO. 2014. Youth and agriculture: Key challenges and concrete solutions. Food and Agriculture Organization of the United Nations (FAO), Technical Centre for Agricultural and Rural Cooperation (CTA), International Fund for Agricultural Development (IFAD), pp.128. https://www.fao.org/3/i3947e/i3947e.pdf

FAO AQUASTAT. 2009. CA Adoption Worldwide, AQUASTAT data query, ©2009 FAO of the UN, Commissioned for the exclusive use of FAO- Conservation Agriculture. http://www.fao.org/ag/ca/6c.html

Flower, K., Crabtree, B. and Butler, G. 2008. No-till Cropping Systems in Australia, In: No-Till Farming Systems. Goddard et al. (eds). World Association of Soil and Water Conservation, Bangkok, Special Publication (3), pp.457-467.

Game, V.N. 2021. Zero Tillage Agricultural Technology. Just Agriculture, pp.42-47.

Govt. of Assam. 2018. Agriculture Policy. Department of Agriculture & Horticulture. P.99. https://agri-horti.assam.gov.in

Gupta, R. K.; Naresh, R. K.; Hobbs, P. R.; Ladha, J. K. 2002. Adopting conservation agriculture in the rice-wheat system of the Indo-Gangetic Plains: New opportunities for saving water. In Water wise rice production. Proceedings of the international workshop on water-wise rice production, eds. Bouman et al. (April 8-11, 2002), pp.207-222. Los Banos, Philippines, International Rice Research Institute.

Hill, M. J., Greaves, H. M., Sayer, C. D., Hassall, C., Milin, M., Milner, V. S., Marazzi, L., Hall, R., Harper, L. R., Thornhill, I., Walton, R., Biggs, J., Ewald, N., Law, A., Willby, N., White, J. C., Briers, R. A., Mathers, K. L., Jeffries, M. J. and Wood, P. J. 2021. Pond ecology and conservation: Research priorities and knowledge gaps. Ecosphere, 12(12), p.22. https://doi.org/10.1002/ecs2.3853

Hobbs, P. R., Giri, G. S. and Grace, P. 1997. Reduced and zero tillage options for the establishment of wheat after rice in South Asia. Rice-Wheat Consortium Paper Series 2. New Delhi, India: RWC.

Hussan, A., Adhikari, S., Das, A., Hoque, F. and Pillai, B.R. 2019. Fish culture without water. International Journal of Fisheries and Aquatic Studies, 7(5), pp. 39-48.

Kumar, R. S., Kundu, S., Kundu, B., Binu, N.K. and Shaji M. 2021. Emerging typology and framing of climate-resilient agriculture in South Asia. The Impacts of Climate Change; A Comprehensive Study of Physical, Biophysical, Social, and Political Issues: Chapter-10, pp.255-287. https://doi.org/10.1016/B978-0-12-822373-4.00021-5

Kungvankij, P., Chua, T.E., Pudadera, B.J., Tiro, Jr. L.B., Corre, Jr.G., Potestas, I.O., Borlongan, E., Taleon, G.A., Alava and Paw, J. N. 1986. Shrimp culture: Pond design, operation and management. Food and Agriculture Organization of the United Nations (FAO), Network of Aquaculture Centres in Asia (NACA), Regional Lead Centre in the Philippines (RLCP). NACA Training Manual Series No. 2. https://www.fao.org/3/AC210E/AC210E06.htm

Machuka, J. 2024. 15 Reasons to Why Educating Farmers is Important. Synnefa. https://help.synnefa.io/articles/15-reasons-to-why-educating-farmers-is-important#:~:text=1.,use%20of%20fertilizers%20and%20pesticides.

Malone, M. and Polyakov, V. 2019. A physical and social analysis of how variations in no-till conservation practices lead to inaccurate sediment runoff estimations in agricultural watersheds. Progress in Physical Geography: Earth and Environment. https://doi.org/10.1177/0309133319873115

Muruganandam, M. 2012a. Why is aquaculture important in watershed management and rural development? Short Course on Watershed-based fisheries development (Muruganandam et al. eds). Central soil and water conservation research and training institute, Dehradun, Uttarakhand, India. pp. 22-34.

Muruganandam, M. 2012b. Aquaculture considerations for the design of watershed ponds. Short Course on Watershed-based fisheries development (Muruganandam et al. eds). Central soil and water conservation research and training institute, Dehradun, Uttarakhand, India. pp. 159-175.

National Research Council. 1999. Drug Use in Food Animals. The Use of Drugs in Food Animals: Benefits and Risks. Washington (DC): National Academies Press (US). 8, Approaches to Minimizing Antibiotic Use in Food-Animal Production. https://www.ncbi.nlm.nih.gov/books/NBK232568/

NFDB. 2016. Recent Trends in Aquaculture: Biofloc Fish Culture. National Fisheries Development Board Department of Fisheries. Ministry of Fisheries, Animal Husbandry & Dairying, Government of India.

Panigrahi, A, Otta, S. K., Kumaraguru, Vasagam, K. P., Shyne, Anand, P. S., Biju, I. F. and Aravind, R. 2019. Training manual on Biofloc technology for nursery and growout aquaculture, CIBA TM series No. 15, p.172.

Patnaik, S. and Ramaprabhu, T. 1985. Aquatic weed management and optimisation of fish culture operation. Lecture Notes on Composite Fish Culture and its Extension in India. Freshwater Aquaculture Research and Training Centre and Network of Aquaculture Centres in Asia. FAO. https://www.fao.org/3/ac229e/AC229E04.htm

Phillips, R. E. and Phillips, S. M. 1984. Ed., No-tillage Agriculture: Principles and Practices. Kluwer Academic Publishers, Netherland, p. 319.

Pingali, P. L. and Shah, M. 2001. Policy re-directions for sustainable resource use: The rice-wheat cropping system of the Indo-Gangetic Plains. Journal of Crop Production, 3 (2), pp.103-118. https://doi.org/10.1300/J144v03n02_05

Prakash, J. 2024. Cultivating A Brighter Future: The Importance of Agricultural Education and Training. The CEO Magazine. https://www.theceo.in/blogs/cultivating-a-brighter-future-the-importance-of-agricultural-education-and-training

Qureshi, A. S., Shah, T. and Akhtar, M. 2003. The groundwater economy of Pakistan. IWMI Working Paper 64. Pakistan Country Series No. 19. Lahore, Pakistan: International Water Management Institute.

Reicosky, D.C. and Saxton, K.E. 1996. Reduced environmental emissions and carbon sequestration. Reprinted 2002. Baker, C.J., Saxton, K.E., Ritchie, W.R., Chamen, W.C.T., Reicosky, D. C., Ribeiro, F., Justice, S.E. and Hobbs, P. R. No-tillage Seeding in Conservation Agriculture. C.J. Baker and K E Saxton (ed.). CABI Publishing. CAB International, Wallingford, UK.

SAFDC (Southeast Asian Fisheries Development Center). 1980. Fish farming handbook. Tigbauan, Iloilo, Philippines: SEAFDEC Aquaculture Department. http://hdl.handle.net/10862/1490

Savage, S. 2017. Is Organic Farming Better for the Environment? Genetic Literacy Project. Retrieved on January 2nd, 2024. https://geneticliteracyproject.org/2017/02/16/organic-farming-better-environment/

Shinde, S.V., and Sukhdhane, K. 2023. Commercial Applications of Biotechnology in Fisheries and Aquaculture. Chronicle of Aquatic Science, 1(1), pp.16- 22.

Tumma, M.K. and Chandrika, K.S.V.P. 2023. Zero tillage Zero worries. Vikaspedia. Accessed on January 10th, 2024. https://vikaspedia.in/agriculture/best-practices/sustainable-agriculture/crop-management/201czero-tillage201d-zero-worries

Ujoh, F.T., Ujoh, F. and Kile, I. 2016. Integrated Production of Rice and Fish: Toward a Sustainable Agricultural Approach. Journal of Scientific Research & Reports, 10(6), pp.1-9.

Vigyan Ashram. 2020. Webinar series: Hydroponics & Aquaponics farming. Vigyan Ashram blog, p.115. https://www.vigyanashram.blog

Vincent C.L., Peigné, J., Casagrande, M. and Silva, E.M. 2017. Overview of Organic Cover Crop-Based No-Tillage Technique in Europe: Farmers' Practices and Research Challenges. Agriculture, 7(5), p.42. https://doi.org/10.3390/agriculture7050042

Wikipedia. 2023. No-till farming. https://en.wikipedia.org/wiki/No-till_farming

Yang, X., Deng, X. and Zhang, A. 2023. Does conservation tillage adoption improve farmers' agricultural income? A case study of the rice and fish co-cultivation system in Jianghan Plain, China. Journal of Rural Studies, 103, 103108. https://doi.org/10.1016/j.jrurstud.2023.103108

Yildiz, H. Y., Robaina, L., Pirhonen, J., Mente, E., Domínguez, D. and Parisi, G. 2016. Fish Welfare in Aquaponic Systems: Its Relation to Water Quality with an Emphasis on Feed and Faeces-A Review. Water, 9(1), p.13. https://doi.org/10.3390/w9010013

13

Analysing the Therapeutic Impact of Marine Organisms: Derived from Osteoporosis Using Machine Learning

G. Chelladurai[1] M. Balasubramanian[2] and J. Nelson Samuel Jebastin[3]

[1]Assistant Professor, Department of Zoology, G.Venkataswamy Naidu College (Autonomous), Kovilpatti-628502, Tamil Nadu, India

[2]Associate professor, Department of Biotechnology, Vivekananda Arts and Science College for Women. Veerachipalayam-637303 Tamil Nadu, India

[3]Assistant Professor, Department of Bioinformatics, Annamalai Univerisity Tamil Nadu-608002, India

---

Abstract

In this nested case-control study, we investigated the impact of VEGFA polymorphisms on the development of bisphosphonate-related osteonecrosis of the jaw (BRONJ) in women with osteoporosis. Our study included 125 individuals and examined 11 VEGFA single nucleotide polymorphisms (SNPs). To identify BRONJ risk factors, we utilized various statistical techniques, such as logistic regression, five-fold cross-validated multivariate logistic regression, elastic net, random forest, and support vector machine models. We evaluated the clinical performance of these models using an AUROC analysis. Our findings revealed a significant association between the VEGFA rs881858 polymorphism and BRONJ development. Specifically, individuals carrying the wild-type rs881858 allele had a 6.45 times higher risk (95% CI, 1.69–24.65) of developing BRONJ than variant homozygote carriers, after adjusting for confounding variables. Additionally, carriers of the rs10434 variant homozygote (GG) showed an increased susceptibility to BRONJ compared to wild-type allele carriers, with an odds ratio of 3.16. Age > 65 years (OR 16.05) and bisphosphonate exposure > 36 months (OR 3.67) were also identified as significant risk factors. All machine learning algorithms used had AUROC scores above 0.78, confirming the correlation between VEGFA polymorphisms and BRONJ incidence in women with osteoporosis.

Keywords: *Marine drug, Osteoporosis, immunosuppressants, Machine Learning*

1. Introduction

Bisphosphonates are frequently utilized in treating osteoporosis and cancer-induced bone metastases, but they carry the risk of inducing a severe complication known as osteonecrosis of the jaw (ONJ). ONJ manifests as exposed bone in the craniofacial area and persists for over eight weeks in patients who are currently or previously treated with bisphosphonates. Many novel antiresorptive tyrosine kinases, mammalian targets of rapamycin, monoclonal antibodies, and immunosuppressant drugs have been discovered since BRONJ was first identified in 2003 and linked to ONJ (Akita *et al.*, 2020). Current ideas include the prevention of angiogenesis and suppression of bone remodeling, gingival inflammation, gingival fibroblast dysfunction, and compromised immunological function. Wound healing mechanisms may play a role in most instances of ONJ, which occur after dental procedures, such as tooth extraction. In addition to surgical treatment for bone lesions, complementary therapies such as laser therapy, ozone therapy, and the administration of platelet concentrates in solid and liquid forms might help prevent ONJ and promote recovery (Gaurav Kumar *et al.*, 2017; Hampton *et al.*, 2015). Formation of new blood vessels is critical for the initiation and maintenance of wound healing.

2. Therapeutic Impact of Marine Organisms

Bisphosphonates are widely prescribed for treating bone disorders such as osteoporosis and cancer-related bone metastases. However, they can lead to a rare but severe adverse effect known as osteonecrosis of the jaw (ONJ). ONJ is characterized by exposed bone in the maxillofacial region persisting for more than eight weeks in individuals who have received current or previous bisphosphonate therapy without prior head and neck radiation treatment. Since its initial identification in 2003, ONJ has been associated with various medications including denosumab, tyrosine kinase inhibitors, mammalian target of rapamycin inhibitors, monoclonal antibodies, radiopharmaceuticals, selective estrogen receptor modulators, and immunosuppressants. Despite extensive research, its pathogenesis remains poorly understood, with proposed mechanisms including suppression of bone remodeling, inflammation, altered gingival fibroblast function, impaired immune response, and inhibition of angiogenesis (Hörner *et al.*, 2020).

Most cases of ONJ occur following dental procedures such as tooth extraction, implicating wound healing processes. Complementary treatments, such as laser therapy, ozone therapy, and platelet concentrate application, in solid and liquid forms, may aid in preventing ONJ and enhancing healing after bone lesion surgery. The effect of bisphosphonates on angiogenic gene expression in healing tissues is hypothesized to contribute to the development of ONJ, given their presumed anti-angiogenic effects. Vascular endothelial growth factor A (VEGF-A) plays a crucial role in wound healing by promoting the proliferation and migration of endothelial cells. Reduced levels of VEGF-A are often observed in patients with ONJ, and bisphosphonates are known to suppress angiogenesis after tooth extraction. Furthermore, microvascular defects have been associated with BRONJ lesions. However, the exact pathophysiology remains unclear, leading to uncertainties regarding the genetic factors involved. Various genes have been implicated in BRONJ, and genome-wide association studies and SNP analyses have identified associations with CYP2C8, PPARG, RBMS4, ASRD, SLC25A5, CCNYL2, SIRT1, FDPS, HLA-DRB1, HLA-DQB1, CYP19A1, and VEGF. Nonetheless, many studies either lack bisphosphonate-exposed controls or focus solely on zoledronic acid use in oncology patients, neglecting patients with osteoporosis.

Although machine learning has been extensively applied in medical fields, including dentistry, its use in predicting BRONJ has not yet been explored. Thus, we aimed to investigate the association between VEGFA polymorphisms and bisphosphonate-related ONJ in patients with osteoporosis and employed supervised machine learning to develop predictive models for BRONJ occurrence.

3. Materials and Methods

3.1 Patients and Data Collection

This prospective, nested case-control study was conducted at Ewha Woman's University Mokdong Hospital between January 2014 and December 2018. Patients scheduled for dentoalveolar surgery who were currently or previously using bisphosphonates were recruited. Eligible participants were individuals aged over 50 years diagnosed with osteoporosis by a medical professional, with exclusion criteria including any history of head and neck radiation or antiresorptive drug administration for tumor treatment. BRONJ was diagnosed by oral surgeons following the guidelines of the American Association of Oral and Maxillofacial Surgeons. This study was approved by the institutional review board of Ewha Woman's University Mokdong Hospital and adhered to the principles of the Declaration of Helsinki. Informed consent was obtained from all participants prior to their inclusion in the study. Clinical data, including age, sex, comorbidities, and duration of bisphosphonate use, were collected from electronic medical records..

3.2 Genotyping

Saliva samples were collected for genotyping using the Oragene® DNA Self-Collection Kit (DNAgenotek, Kanata, ON, Canada) in tube format (OG300). The samples were incubated two-hour incubation at 50°C before DNA extraction according to the manufacturer's protocol. Genomic DNA was extracted according to the manufacturer's instructions. Genetic data for VEGFA SNPs were obtained from Haploreg 4.1, the SNP database of the National Center for Biotechnology Information (NCBI). Eleven VEGFA SNPs (rs2010963, rs699947, rs10434, rs25648, rs3024987, rs3025022, rs3025035, rs3025039, rs998584, rs6905288, and rs881858) were selected and genotyped to investigate their associations with BRONJ development. Genotyping of these SNPs was conducted using SNaPShot Multiplex kits (ABI, Foster City, CA, USA) or TaqMan genotyping assays with a real-time polymerase chain reaction system (ABI 7300, ABI) following the manufacturer's instructions.

3.3 Statistical Analysis and Machine Learning Methods

The chi-squared test was used to compare categorical variables, while Student's t-test was used to compare continuous variables between the case and control groups. Multivariate logistic regression analysis determined independent risk factors for BRONJ, including elements with p values < 0.05, in univariate analysis. Odds ratios (ORs) and adjusted odds ratios (ORs) were calculated for both univariate and multivariate analyses. Machine learning algorithms have been developed to predict BRONJ risk factors, including five-fold cross-validated multivariable logistic regression, elastic net, random forest (RF), and support vector machine (SVM) classification models, implemented using the R package caret. The dataset was randomly divided into five subsets for cross-validation, with one subset used for model validation, and the remainder for model training. Each cross-validation iteration was repeated 100 times to assess the model performance. In the elastic net, the grid-search values for λ and α were varied to optimize the penalty weight and the importance of the ridge and lasso penalties, respectively. For the RF, the number of randomly selected predictors (entries) was tested. For SVM, linear and radial kernel functions were utilized, with optimization of the cost and sigma parameters. Model performance for BRONJ occurrence was evaluated using the area under the receiver operating curve (AUROC) and 95% confidence interval (CI) for each model. All statistical analyses used a two-tailed alpha of 0.05. Data analysis was conducted using the Statistical Package for Social Sciences Version 20.0 for Windows (SPSS, Chicago, IL, USA), while R software version 3.6.0 was employed for constructing machine learning algorithms.

4. Results

The study encompassed 149 patients initially, but 24 were excluded for various reasons: 20 due to additional indications beyond osteoporosis, two lacking clinical information, and two being male. The remaining 125 patients were included in the final analyses. Notably, 58 patients (46.4%) developed bisphosphonate-related osteonecrosis of the jaw (BRONJ) following dental procedures. The mean age of the participants was 72.9 ± 9.4 years, with 19 patients aged < 65 years. Hypertension was more prevalent among the cases than among the controls (62.4% vs. 41.8%, $p < 0.05$). Additionally, the proportion of patients treated for 36 months or longer was significantly higher in the case group than in the control group ($P < 0.01$). Regarding genetic factors, among the 11 VEGFA SNPs studied, all allele frequencies were aligned with Hardy–Weinberg equilibrium. Univariate analysis identified that rs10434 (A > G) and rs881858 (G > A) were significantly associated with BRONJ development. Notably, variant homozygous carriers (GG) of rs10434 had a higher incidence of BRONJ than other genotypes, while carriers of the wild G allele of rs881858 faced an elevated risk of BRONJ development. Adjusting for demographic variables with $p < 0.05$ revealed that the odds of BRONJ development were approximately 6.45 times higher among G allele carriers of rs881858 compared to variant homozygote carriers ($p < 0.01$). Although the association with rs10434 reached marginal significance after covariate adjustment, patients with the GG rs10434 genotype showed a notably increased risk of BRONJ. In terms of demographic factors, patients treated for more than 36 months and those older than 65 years had higher odds ratios for BRONJ occurrence. The attributable risk for rs881858 polymorphism was substantial (84.5 %). Machine learning models, including logistic regression, elastic net, random forest (RF), and support vector machine (SVM), demonstrated favorable performance in predicting BRONJ occurrence, with an average AUROC value of 0.78 across 100 random iterations.V EGF plays a crucial role in vascular development and angiogenesis by stimulating endothelial cell proliferation and increasing vascular permeability. The VEGF protein family includes VEGF-A (also known as VEGF), VEGF-B, VEGF-C, and VEGF-D, which exert their effects by binding to VEGF receptors, such as VEGFR1, VEGFR2, and VEGFR3. The VEGF-A/VEGFR2 pathway is a well-known inducer of angiogenesis that influences various signalling pathways during vascular formation. The interplay between various molecular pathways, particularly those involving vascular endothelial growth factor (VEGF), plays a critical role in endothelial proliferation, angiogenesis, and subsequently, bone repair and wound healing processes. Long-term bisphosphonate therapy, while effective for osteoclast suppression, can detrimentally affect osteoblast function, thereby hindering bone repair. Studies in mice with osteoblast-specific deletion of VEGFA have highlighted the necessity of appropriate VEGF levels to couple angiogenesis with osteogenesis during repair. Disruption of VEGF signaling, as observed with bisphosphonate therapy, can impede bone healing by interfering with the conversion of cartilage calli to bone calli, potentially leading to avascular necrosis and impaired tissue repair. The rs881858 SNP, a significant factor in our study, is located in a VEGFA intron. Although its specific function remains uncharacterized, it may influence VEGFA gene activity regulation through chromosomal interactions with the VEGFA promoter region. Prior genomic investigations have associated wild-type (G allele) rs881858 with conditions such as chronic kidney disease and insulin resistance, suggesting a role in angiogenesis impairment. Similarly, the rs10434 SNP, located in the 3′UTR region of VEGFA, has been implicated in various diseases, such as chronic lymphocytic leukemia and pre-eclampsia, possibly via modulation of VEGFA expression. While numerous potential risk factors for bisphosphonate-related osteonecrosis of the jaw (BRONJ) have been proposed, evidence remains limited due to the lack of prospective studies. Nonetheless, in our study, age and treatment duration were significant demographic risk factors for BRONJ, which is consistent with previous findings. In predicting BRONJ occurrence, diverse machine learning approaches were employed, all of which demonstrated robust performance, as indicated by the AUROC values. Notably, the multivariable logistic regression model and elastic net model exhibited comparable predictive accuracy, confirming the utility of these methods in risk assessment and management. Overall, these findings underscore the intricate interplay among genetic factors, environmental influences, and treatment characteristics in BRONJ pathogenesis, emphasizing the importance of comprehensive risk stratification and management strategies in clinical practice. The penalized linear regression model utilized in this study combines the lasso and ridge methods, offering a balance between variable selection and shrinkage. Random Forest (RF) is an ensemble method that constructs multiple decision trees based on bootstrapped samples of the training data, employing a random subset of features at each node for optimal splitting. The RF provides robust and reliable classification results by aggregating the predictions of these trees. Support Vector Machines (SVMs) were employed with both linear and radial basis function kernels. Linear kernel SVMs utilize a single tuning parameter, C, which controls the trade-off between model complexity and error, whereas radial kernel SVMs incorporate an additional hyperparameter defining the spread of the Gaussian kernel.

Despite its contributions, this study had several limitations. The sample size was small, which led to underpowered analyses. Furthermore, the inclusion of only two male participants precluded gender-based analyses. Covariates, such as smoking history and corticosteroid therapy, were inadequately addressed due to insufficient information, potentially introducing confounding effects. Moreover, certain unmeasured confounders may have been overlooked in the predictive models. Additionally, the study did not investigate the molecular mechanisms underlying the development of BRONJ. The lack of external validation and consideration of other factors influencing the machine learning algorithm performance further underscores the need for cautious interpretation. Nevertheless, this study is the first to employ machine learning methods for BRONJ prediction. The control group, comprising patients carefully selected by oral and maxillofacial surgeons post-dentoalveolar surgery, enhanced the credibility of the study by minimizing bias often associated with including healthy or ambiguous controls in genetic research.

Conclusion

To the best of our knowledge, this is the first study to examine the impact of VEGFA gene polymorphisms on BRONJ occurrence among osteoporotic patients. Additionally, our utilization of advanced machine-learning techniques to predict the probability of BRONJ adds further novelty to our approach. Although we recognize the necessity for additional functional investigations to validate our findings, our results hold promise for providing valuable insights for clinical decision-making concerning ONJ risk assessment.

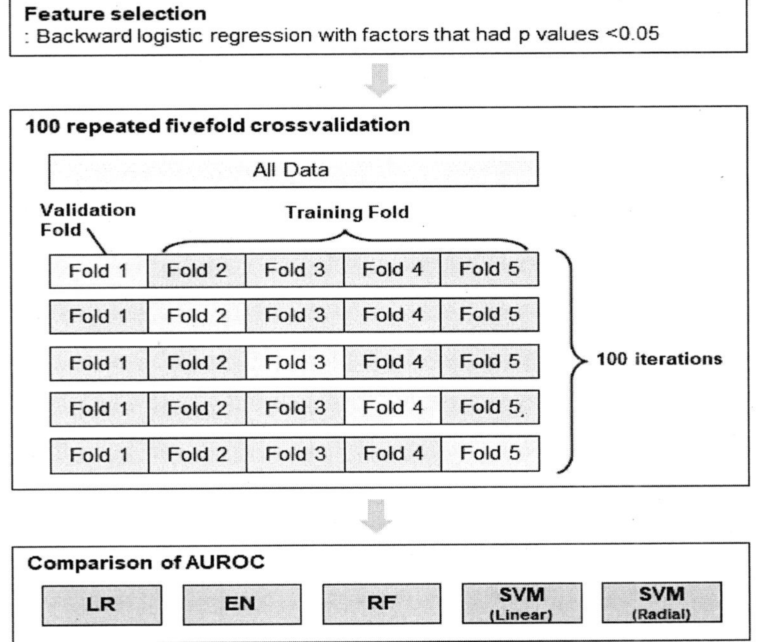

Fig. 1. Depicts the Flow chart of the machine learning approaches.

Table 1. Depicts the Patient characteristics of the study patients.

Gene Polymorphism		Minor Allele Frequency	Groupea Genotypes	Case (n=58)	Control (n=67)	p
rs699947	A > C	0.253	AA, AC	22 (37.9)	31 (46.3)	0.347
			CC	36 (62.1)	36 (53.7)	
rs2010963	C > G	0.439	CC	14 (25.0)	8 (125)	0.077
			CG, GG	42 (75.0)	56 (87.5)	
rs25648	C > T	0.081	CC	51 (87.9)	52 (77.6)	0.131
			CT, TT	7 (12.1)	15 (22.4)	
rs3024987	C > T	0.211	CC, CT	56 (96.6)	63 (94.0)	0.685
			TT	2 (3.4)	4 (6.0)	
rs3025022	C > T	0.181	CC, TT	18 (31.0)	23 (34.3)	0.696
			TT	40 (69.0)	44 (65.7)	
rs3025035	C > T	0.202	CC	34 (59.6)	49 (73.1)	0.246
			CT, TT	23 (40.4)	18 (26.9)	
rs3025039	C > T	0.133	CC	42 (72.4)	50 (74.6)	1.000
			CT, TT	16 (27.6)	17 (25.4)	
rs10434	A > G	0.113	AA, AG	7 (12.1)	18 (26.9)	0.039
			GG	51 (87.9)	49 (73.1)	
rs998584	C > A	0.421	CC	7 (12.1)	14 (21.1)	0.176
			CA, AA	51 (87.9)	52 (78.8)	
rs6905288	G > A	0.240	GG, GA	21 (36.2)	33 (49.3)	0.142
			AA	37 (63.8)	34 (50.7)	
rs881858	G > A	0.133	GG, GA	18 (31.0)	10 (14.9)	0.031
			AA	40 (69.0)	57 (85.1)	

Table 2. Depicts the Associations of genotypes with bisphosphonate-related osteonecrosis of the jaw

Characteristics	**Case (n=58)**	**Control (n=67)**	**p**
Age (years)			0.003
<65	3 (5.2)	16 (24.2)	
≥65	55 (94.8)	50 (75.8)	
Comorbidity			
Hypertension	36 (62.1)	28 (41.8)	0.024
Diabetes Mellitus	18 (31.0)	16 (23.9)	0.370
Cardiovascular Disease	8 (13.8)	8 (11.9)	0.757
Rheumatoid Arthritis	7 (12.1)	2 (3.0)	0.080
Thyroid Disease	4 (6.9)	2 (3.0)	0.415
Kidney Disease	2 (3.4)	3 (4.5)	1.000
Liver Disease	0 (0)	2 (3.0)	0.499
Cancer	2 (3.5)	6 (9.1)	0.284
Treatment Duration (Months)			
<36	13 (25.5)	30 (55.6)	0.002
≥36	38 (74.5)	24 (44.4)	

Table 3. Depicts the Multivariate analysis to identify predictors of bisphosphonate-related osteonecrosis of the jaw.

Variables	**Crude Odds Ratio (95% CI)**	**Adjusted odds Ratio (95% CI)**	**Attributable Risk (%)**
Age ≥ 65 years	5.87 (1.61-21.34)**	16.05 (1.87-138.05)*	93.8
Treatment Duration ≥ 36 month is	3.65 (1.60-8.36)**	3.67 (1.36-9.94)*	72.8
VEGFA			
rs10434, GG	2.68 (1.03-6.97)*	3.16 (0.97-1031)	68.4
rs881858, GG/GA	2.56 (1.07-6.14)*	6.45 (1.69-24.65)**	84.5

References

Akita, M., Nishikawa, Y., Shigenobu, Y., Ambe, D., Morita, T., Morioka, K., & Adachi, K. (2020). Correlation of proline, hydroxyproline and serine content, denaturation temperature and circular dichroism analysis of type I collagen with the physiological temperature of marine teleosts. Food Chemistry, 329, 126775. https://doi.org/10.1016/j.foodchem.2020.126775

Gaurav Kumar, P., Nidheesh, T., Govindaraju, K., Jyoti, S. P. V., & Suresh, P. V. (2017). Enzymatic extraction and characterisation of a thermostable collagen from swim bladder of rohu (Labeo rohita). Journal of the Science of Food and Agriculture, 97(5), 1451–1458. https://doi.org/10.1002/jsfa.7884

Hampton, A. L., Aslam, M. N., Naik, M. K., Bergin, I. L., Allen, R. M., Craig, R. A. *et al.* (2015). Journal of the American Association for Laboratory Animal Science, 54/5, 487–496.

Hörner, C., Schürmann, C., Auste, A., Ebenig, A., Muraleedharan, S., Dinnon, K. H., Scholz, T., Herrmann, M., Schnierle, B. S., Baric, R. S., & Mühlebach, M. D. (2020). A highly immunogenic and effective measles virus-based Th1-biased COVID-19 vaccine. Proceedings of the National Academy of Sciences of the United States of America, 117(51), 32657–32666. https://doi.org/10.1073/pnas.2014468117

Lacativa, P. G., & Farias, M. L. (2010). Osteoporosis and inflammation. Arquivos Brasileiros de Endocrinologia e Metabologia, 54(2), 123–132. https://doi.org/10.1590/s0004-27302010000200007

O'Gorman, D. M., Tierney, C. M., Brennan, O., & O'Brien, F. J. (2012). The marine-derived, multi-mineral formula, Aquamin, enhances mineralisation of osteoblast cells in vitro. Phytotherapy Research, 26(3), 375–380. https://doi.org/10.1002/ptr.3561

Pal, P., Srivas, P. K., Dadhich, P., Das, B., Maity, P. P., Moulik, D., & Dhara, S. (2016). Accelerating full thickness wound healing using collagen sponge of mrigal fish (Cirrhinus cirrhosus) scale origin. International Journal of Biological Macromolecules 93/B, 93(B), 1507–1518. https://doi.org/10.1016/j.ijbiomac.2016.04.032

Ryan, S., O'Gorman, D. M., & Nolan, Y. M. (2011). Evidence that the marine-derived multi-mineral Aquamin has anti-inflammatory effects on cortical glial-enriched cultures. Phytotherapy Research, 25(5), 765–767. https://doi.org/10.1002/ptr.3309

Slevin, M. M., Allsopp, P. J., Magee, P. J., Bonham, M. P., Naughton, V. R., Strain, J. J., Duffy, M. E., Wallace, J. M., & Mc Sorley, E. M. (2014). Supplementation with calcium and short-chain fructo-oligosaccharides affects markers of bone turnover but not bone mineral density in postmenopausal women. Journal of Nutrition, 144(3), 297–304. https://doi.org/10.3945/jn.113.188144

Terada, M., Izumi, K., Ohnuki, H., Saito, T., Kato, H., Yamamoto, M., Kawano, Y., Nozawa-Inoue, K., Kashiwazaki, H., Ikoma, T., Tanaka, J., & Maeda, T. (2012). Construction and characterization of a tissue-engineered oral mucosa equivalent based on a chitosan-fish scale collagen composite. Journal of Biomedical Materials Research. Part B, Applied Biomaterials, 100(7), 1792–1802. https://doi.org/10.1002/jbm.b.32746

Zhao, R. (2012). Immune regulation of osteoclast function in postmenopausal osteoporosis: A critical interdisciplinary perspective. International Journal of Medical Sciences, 9(9), 825–832. https://doi.org/10.7150/ijms.5180

Zhu, C. F., Li, G. Z., Peng, H. B., Zhang, F., Chen, Y., & Li, Y. (2010). Therapeutic effects of marine collagen peptides on Chinese patients with type 2 diabetes mellitus and primary hypertension. American Journal of the Medical Sciences, 340(5), 360–366. https://doi.org/10.1097/MAJ.0b013e3181edfcf2

14

Carbon and Nitrogen Ratio in Shrimp Aquaculture Systems

H.S. Praveenjoshi[1*]*, K.R. Amogha*[2]*, Narendra Kumar Maurya*[1] *and Shiwam Dubey*[1]

[1]*Department of Aquatic Environment Management, College of Fisheries, Karnataka Veterinary, Animal and Fisheries Sciences University, Mangaluru-575002 Karnataka, India*

[2]*Department of Aquaculture, College of Fisheries, Karnataka Veterinary Animal and Fisheries Sciences University, Mangaluru-575002, Karnataka, India*

Abstract

This abstract discusses the importance of aquaculture and shrimp farming in meeting the global demand for food. This highlights the advancements in technology and growing demand for fish as a food source. This article emphasizes the importance of providing suitable conditions in captivity by maintaining the physicochemical parameters at the optimum level required by the species cultured. It also discusses the challenges in shrimp farming and the need for suitable management practices to make shrimp farming socially, environmentally, and economically viable. This article also provides information on the importance of shrimp species in aquaculture and their cultivation techniques.

1. Aquaculture and Shrimp farming

Considering the current population explosion and the speed of human-centered land encroachment, scientists have expressed concern about the availability and production of food to meet global demand. The agricultural sector alone would not be able to cope with such a huge global demand of 7.6 billion people, which does not seem to be decreasing in any way. Approximately 20% of the global population consumes fish, which originate in the aquaculture sector. On the other hand, it has grown many fold with advancements in science and technologies in fish production and post-harvest, which has become a prominent commercial sector. The fisheries sector is broadly classified into two types: marine and inland. In marine fisheries, fishermen venture into the sea to fish with their fishing vessels and gear to harvest fish, whereas in inland fisheries, fishermen go fishing in rivers, lakes, and reservoirs. However, with advancements in technology and growing demand for fish as food sources, people have started growing fish in captivity under controlled and semi-controlled conditions in ponds. The history of fish culture tracks back to 2000 BC, which originated in China, where they started growing carp in captivity by providing the required conditions for fish growth. Providing suitable conditions in captivity means providing the required physical and chemical environment in the culture system by keeping the physicochemical parameters at the optimum level as required by the species cultured. Physicochemical parameters of soil and water are key to the successful culture of fish, which is why emphasis has been placed on the scientific study of the physical and chemical parameters of soil and water.

Fin/shellfish farming is one of the world's fastest growing aquaculture subsectors. India has a vast coastline of 8,118 km in geographical area and an exclusive economic zone (EEZ) of 2.02 million km2. Socio-economically underdeveloped small and small-scale fishermen, whose lives are closely intertwined with the ocean and sea, who dominate sea fishing, in order to improve fishing resources and replenish natural stocks whose populations have decreased due to overexploitation or environmental degradation, marine fin fish culturing has become more and more popular. Shrimp aquaculture has been used for many years in Southeast Asia, and is a traditional type of coastal farming in several nations. A variety of issues have arisen as a result of the recent tendency toward a more intense culture. Experiences in the area, however, indicate that with the adoption of suitable management practices, shrimp farming can be socially, environmentally, and economically viable and help produce food and reduce poverty in coastal areas. The creation and application of such management strategies must consider the technical, economic, social, and environmental challenges.

According to the NFDB (2020), India ranks triennially in fisheries results and second in aquaculture. Fisheries has working 145 heap nations and provided 1.07% of the GDP and export gain of Rs 334.41 billion. Immediately, Andhra Pradesh tops under idea and the result of insignificant. Marketing of shrimp gardening began in the old age of 2009–2010 (MPEDA, 2021). Total planet fisheries and growing plants in liquid production achieved another extreme record of 178.5 mmt at which point, 96.4 heap tonnes is captured fisheries while 82.1 heap tonnes were from aquaculture results in 2018 (FAO, 2020). According to FAO (2020), at the end of 1961–2017, the average annual tumor rate of total cooking net consumption increased by 3.1%, surpassing the annual study of the human population of 1.6%. The cooking angle consumption per person raised from 9.0 kilograms (live burden equivalent) in 1961 to 20.3 kilograms in 2017. Growing plants in a liquid can control overexploitation, conceive jobs, and specify the experience accompanying protein-rich fares. The Administration of India plans to increase extricate production from 137.58 lakh to 220,000 tonnes in 2018-19 under the Pradhan Mantri Matsya Sampada Yojana (PMMSY) of the Bureau of Fisheries, Farming and Dairying, Management of India.

2. Important Shrimp Species in Aquaculture

Chiefly crustaceans, cougar shrimp (Penaeus monodon), Indian white shrimp (Penaeus indicus), white leg shrimp (Penaeus vannamei), rose tail shrimp (Penaeus penicillatus), banana shrimp (Penaeus merguiensis), green cougar shrimp (Penaeus indicus), an insignificant secondhand in coastal growing plants in liquid. The Peaceful silver insignificant or white move along the foot insignificant Litopenaeus vannamei (Boone, 1931) is the top secret in the Phylum Arthropoda, Class Crustacea, Order Decapoda, Family Penaeidae, Type Litopenaeus, and Variety vannamei (Perez and Kensley, 1997). It is owned by the tropical situated or toward the west Appeasing coast of Concerning ancient culture America, from about south Mexico in the north to Peru on the west side when facing north, between latitudes of 32°N and 23°S. This peneid is very plentiful on the coast of Ecuador in Esmeraldas (border responsibility of Colombia), place females are handy during the whole of old age (Huang *et al.*, 2003). Litopenaeus vannamei is intensely euryhalic, worthy of low salinity waters of 1-2 psu and hypersaline waters of up to 40 psu (Menz and Blake, 1980). Juveniles and juveniles persisted at a dirty bottom in warm water (25–32 °C) with a salinity of 28–34 psu at an insight of 70 cm and a burrow. Persons favor higher salinities of 34-35 psu and favor kind of deeper water (30-50m). The young plethora presented an inverse relationship with accompanying salinity and a definite correlation in a Mexican seaside pond structure (Rivera *et al.* 2008).

Recently, L. vannamei has been acquired by coastal aquaculture farmers, as it is one of the most intensively cultivated shrimps worldwide and has a reduced risk of catastrophic diseases (Perez and Kensley, 1997; Boyd, 2002; Zhu *et al.*, 2006). The intensification of production systems leads to adverse changes in water quality and increases the risk of diseases owing to higher stocking densities and feeding rates (Nasrin, 2016). Shrimp farming has been practiced in India for 40 years; however, commercial and large-scale shrimp culture began in the 1990s.

Originally, insignificant farming was approved only on an exploratory scale in India. A main step towards large-scale insignificant growing plants in liquid was captured soon after the Indian Council of Agricultural Research (ICAR) Central Inland Fisheries Research Institute first manifested salt solution fish gardening in West Bengal in 1973. Subsequently, a research project matched by the ICAR on salt solution aquaculture was distracted. up across India in 1975 in West Bengal, Andhra Pradesh, Odisha, Tamil Nadu, Goa and Kerala. At the same time, the Main Marine Fisheries Research Institute (Vijayan and Kailasam, 2020) reported insignificant seed results at Narakkal, Kerala. Marketing insignificant hatcheries were established for the Sea Device Export Happening Expert. Similarly, almost-intensive cultivation science has been popularized in ship-scale MPEDA (Muralidharan, 2019). As insignificant aquaculture spread throughout India, these methods, in addition to experiments by other growers, allowed the industry to evolve considerably. The Biofloc order was developed to improve the preservation of natural resources for amphibious animal production. In growing plants in liquid, feed costs (60% of total costs) and the availability of water and land are the ultimate limiting determinants. The law concerning this technique identifies a nitrogen cycle by asserting a taller C/N percentage (the ratio of element to nitrogen), exciting the development of heterotrophic microbes that adjust nitrogen-rich waste that may be secondhand as feed by cultivated variety. A greater C/N percentage is maintained by increasing the beginning of hydrogen (molasses), and the water type is enhanced by the presence of high-quality distinct container proteins. The confinement of the toxic nitrogen class occurs faster in biofloc science (BFT) because the microbial result of heterotrophs is 10 times higher than that of autotrophic microorganisms. Because of the lower residence and resistance to tangible changes, this method was adopted in insignificant ranching (Avnimelech, 1999).

3. Physico-Chemical Characteristics of Water and Soil in Shrimp Ponds

Understanding the physicochemical characteristics of water and soil in shrimp ponds and the growth performance of aquatic animals is necessary to control the disease and prevent stress in shrimp. At present, L. vannamei is the most preferred species for shrimp producers due to its short crop time, hardy species, and high market value. Water quality plays a vital role in maintaining aquatic animal health, growth performance, and survival rate in shrimp ponds. Due to increased stocking density, feeding rate, and pollutant intake, water faces the risk of water quality issues being common in shrimp ponds, but management of the aquatic environment is a challenge. Poor water quality causes diseases, mortality, slow growth, and low shrimp production. However, the shrimp aquaculture industry has faced various problems, such as germplasm degradation, disease outbreaks, and water quality degradation, which have seriously hindered its further development (Bachere, 2000; Thitamadee *et al.*, 2016). In addition, the high-water turnover in the aquaculture process not only causes a loss of nutrients but also seriously pollutes the surrounding environment (Bachere, 2000).

In intensive aquaculture systems, waste consisting mostly of excrement and unused feed leads to the accumulation of toxic metabolites such as ammonium and nitrite, which degrade shrimp habitats (Avnimelech and Ritvo, 2003; Piedrahita, 2003). Increasing the C/N ratio to promote heterotrophic microbes in pond sewage systems regulates water quality by removing toxic nitrogen, such as ammonia. Heterotrophic microbes are mainly responsible for performing the necessary functions in the biofloc system, and their ammonium fixation by heterotrophic bacteria occurs much faster than nitrifying bacteria due to their faster growth rate and higher yield of microbial biomass per substrate unit. The bacterial protein produced by the assimilation of ammonium nitrogen must have a sufficient content of protein, lipids, carbohydrates, and ash as a high-quality aquaculture feed. As organic matter decomposes, carbon breaks down faster than nitrogen, thereby reducing the ratio of carbon to nitrogen. Adding compost or other nutrients can help to determine the correct carbon-nitrogen ratios. A large amount of raw organic matter can be applied to soil, thereby microorganisms multiply rapidly, but in the process of operation, they consume nitrogen. Dissolved oxygen (DO) is one of the most critical water quality parameters for aquaculture monitoring. Biofloc technology is defined as "the use of matrix-perpetuating aggregates of bacteria, algae, or protozoa with particulate organic matter to improve water quality, waste management, and disease prevention in intensive aquaculture systems" (Avnimelech, 1999). In addition to the oxygen demand of farmed shrimp or fish, the rich microbial community consumes a significant extent. The intensity of the DO consumption by the microbial community is highly dependent on the feed required for a given stocking density (Boyd, 2009).

The microbial society arranges the reuse of excess foods. In specific plans, pieces are frequently removed by outside filtration to a certain degree of sedimentation, whirlpool, and soil filters. However, in biofloc systems, pieces are permitted to form in a breeding scheme, and the microbial community is unspecified because the mineral controlling a vehicle is inside these pieces. The main principle of element and nitrogen search weakens the water exchange and spurs the development of heterotrophic structures utilizing the nitrogen waste involved in the pool or container. The C/N ratio is continually diminished in biofloc growing plants in the liquid method by increasing organic elements as a beginning to the animal container that does not use organic nitrogen for microbial aggregation. An equalized percentage of elemental nitrogen in the feed is the main factor that exaggerates the growth of heterotrophic microorganisms. This process was used to manage the nitrogen content of the water. The ultimate standard C/N percentages are 10:1 and 15:1, which means that, at a C/N ratio of 10:1, 10 element beginnings are required to destroy individual nitrogen. In growing plants in liquid, the C/N percentage is deliberately established as the protein content of the feed and total ammonium nitrogen (TAN). When carbon and nitrogen are in a balanced liquid, ammonium is added to bacterial biomass apart from basic nitrogen-containing wastes (Schneider *et al.*, 2005). In populated growing plants in liquid ponds, water administration is paramount. The same water status impairs development and endurance. Good water quality normally refers to the rightness of the water for insignificant continuation and growth. By increasing carbohydrates in the pool, tumors of heterotrophic microorganisms are aroused, and nitrogen adjustment occurs as a result of microbial proteins (Avnimelech, 1999).

4. Carbon and Nitrogen ratio (C/N ratio)

The element-to-nitrogen percentage (C/N ratio) is the process of determining the amount of nitrogen in water. Most standard C/N ratios are 10:1 and 15:1, which means that at a C/N ratio of 10:1, 10 element beginnings are required to kill 1 Nitrogen. Avnimelech (1999) gives welcome views that, in growing plants in liquid, C/N percentage has been determined by established protein allotment of feed and total liquid nitrogen (TAN). If carbon and nitrogen are present in the solution, ammonium, in addition to basic nitrogenous waste, is converted into bacterial biomass (Schneider *et al.*, 2005). In exhaustive growing plants in liquid systems, waste created throughout the course of civilization, primarily feces and remaining feed, encourages the accretion of poisonous metabolites such as ammonium and nitrite, thereby babying the living surroundings of the insignificant (Avnimelech and Ritvo, 2003). In populated aquaculture ponds, water administration is the principal. The depth of the water quality impairs the tumor and endurance. Good water characteristics are normally the condition or suitability of water for insignificant continuation and tumors. By adding carbohydrates to the pool, the progress of heterotrophic microorganisms is increased, and nitrogen absorption occurs as a result of microbial proteins (Avnimelech, 1999). Wujie *et al.*, (2016) examined the effect of the C/N ratio on biofloc development and afterwater characteristics and adeptness of L. vannamei cooling in biofloc-based extreme-bulk nothing-exchange outside tank plans. They stated that the changeable suspended clump (VSS) and turbidity principles were better determinable limits for biofloc quantification than the postponed mass of material (SS) or total postponed chunk (TSS). TAN and NO2 concentrations can be efficiently used for conditional heterotrophic absorption. Autotrophic nitrification helps claim shrimp concentrations within satisfactory ranges for insignificant agriculture even at high sock densities. Microalgae and autotrophic microorganisms are more advantageous than heterotrophic microorganisms for insignificant performance in extreme-mass nothing-transformation breeding structures. Muthusamy *et al.*, (2016) reported a decline in liquid total nitrogen content while asserting good water features in shrimp sophistication in biofloc sophistication orders. Water value and shrimp results were recorded in widely governed ponds with or outside a hydrogen-located diet for a very small person (Hari *et al.*, 2006). Jaganmohan and Leela (2018) stated that all the proven limits such as pH, Salinity, Carbonates, Bicarbonates, Total alkalinity, Calcium, Magnesium, Total severity, Total liquid and Nitrite upheld under optimum conditions were appropriate for L. vannamei production. This was following Islam and others. (2004), salinity fluctuated from 3.0 to 15.0 ppt in the southwest, when in fact it was between 2.5 and 20.0 ppt in the southeast domain, and total liquid-nitrogen higher than the urged level for insignificant breeding in Bangladesh. Claude and Gross (2014) intentionally used probiotics to improve soil and water characteristics in growing plants in liquid ponds and reported very few beneficial benefits of probiotics on water and bottom soil quality.

Panigrahia and others. (2018) intentional Carbon: Nitrogen (C:N) percentage affected the microbial community of bureaucracy and tumor, in addition to immunity of insignificant (L. vannamei) in biofloc-located culture structures. The flow of Vibrio supremacy decreased with an increase in C: N percentage, thus ratifying the supremacy of heterotrophic microorganisms in the high C: N percentage groups. Upon challenge with the accompanying pathogens, shrimp from the C: N10, C: N15, and C: N20 groups showed considerably greater survival ($p<0.05$) than those from the C: N5 and control groups. Intentional performance of Platymonas and Microbial Society under different C/N percentages in biofloc science-growing plants in liquid systems. Platymonas sp. and C/N percentages considerably affected the variety and stock copiousness. Platymonas sp. and related microorganisms in bioflocs have an advantageous effect on water quality by lowering nitrogen compounds, providing a good environment for certain groups of microorganisms, and lowering dependence on greater concentrations of element beginnings. Fontenot *et al.* (2007) stated the effect of hotness, salinity and carbon:nitrogen percentage on a subsequent catalyst treating insignificant growing plants in liquid wastewater. The results showed that a salinity of 28-40 ppt, hotness range of 22-37oC, and C: N ratio of 10:1 presented the best results in terms of maximum nitrogen and element removal from wastewater. Asaduzzaman *et al.*, (2010), C/N percentage and substrate addition in natural trophic societies in insignificant freshwater ponds increased the C/N percentage considerably and guided the biomass of plankton, periphyton, heterotrophic bacteria, and benthic macroinvertebrates. However, basin societies are underutilized by freshwater shrimps. Therefore, it is necessary to further investigate the chance of lowering the affected feeding or increasing the insignificant stocking bulk. Chakrapani *et al.*, (2020) experimented with Pacific silver insignificant Penaeus vannamei with three different C: N percentages under useful conditions that considered growth effectiveness, invulnerable reactions, and metabolic pathways. However, the results demonstrated that insignificant growth in C/N 10 (630 mg) and C/N 15 (646 mg) biofloc structures occurred considerably faster than in the C/N 20 (528 mg) and control groups (374 mg). Concerning extracellular enzymes, proteases, lipases, and xylanases were expected to be mostly present in settled microorganisms isolated from biofloc situations, whereas amylase was usually present in all the treatments.

Conclusion

The Carbon and Nitrogen ratio is critically important in shrimp aquaculture due to its impact on various aspects of system health, efficiency, and shrimp well-being. The key reasons why managing the C:N ratio is crucial, A balanced ratio supports beneficial microbial communities that break down organic matter. If the ratio is off, microbial processes can become inefficient, leading to poor water quality and also it facilitate the decomposition of

organic waste, reducing the accumulation of sludge and harmful byproducts like ammonia and nitrites, which can be toxic to shrimp. An optimal C: N ratio ensures that nitrogen from feed and waste is used effectively by microbes for breaking down organic matter. This balance helps prevent nutrient imbalances in the system, which can lead to algae blooms and other water quality issues and helps in helps in optimizing feed formulations. Excess carbon or nitrogen can lead to inefficient feed utilization and increased waste, affecting growth rates and overall shrimp health. Stable water quality and efficient waste management supported by a proper C: N ratio contribute to better growth rates and feed conversion efficiency in shrimp and Balanced nutrient levels and clean water reduce the risk of diseases and stress, leading to healthier shrimp and improved survival rates. Efficient nutrient management minimizes environmental impacts by reducing waste outputs and preventing pollution, aligning with sustainable aquaculture practices. C:N ratio is a fundamental parameter in shrimp aquaculture that influences microbial activity, water quality, nutrient management, shrimp health, and system efficiency. Properly managing this ratio is essential for successful and sustainable aquaculture operations.

References

Asaduzzaman, M., Rahman, M.M., Azim, M.E., Ashraful, I.M., Wahab, M.A., Verdegem, M.C.J. And Verreth, J.A.J., (2010). Effects of C/N ratio and substrate addition on natural food communities in freshwater prawn monoculture ponds. Aquaculture, 306:127–136. http://dx.doi.org/10.1016/j.aquaculture.2010.05.035

Avnimelech, Y., (1999). Carbon/nitrogen ratio as a control element in aquaculture systems. Aquaculture, 176: 227–235.

Avnimelech, Y. And Ritvo, G., (2003). Shrimp and Fish Pond Soils: Processes and Management. Aquaculture, 220: 549-567. doi:10.1016/S0044-8486(02)00641-5

Bachere, E., (2000). Shrimp immunity and disease control. Aquaculture, 191: 3–11. https://doi.org/10.1016/S0044-8486(00)00413-0

Boyd, C.E. (1995). Pond Bottom Soil Analyses. In: Bottom Soils, Sediment, and Pond Aquaculture. Springer, Boston, MA. https://doi.org/10.1007/978-1-4615-1785-6_10

Boyd, C.E. (2002). Standardize terminology for low salinity shrimp culture. Global Aquaculture Advocate, 5(5): 58-59.

Boyd, C.E., (2009). Pond bottom soil analyses. Global Aquaculture Advocate, September and October. USA, 92 pp.

Chakrapani, S., Akshaya, P., Jayashree, S., Mullaivanam, R., Sivakumar R.P. And Vinodh, K., (2020). Three different C: N ratios for Pacific white shrimp, Penaeus vannamei under practical conditions: Evaluation of growth performance, immune and metabolic pathways. Aquaculture Research, 1-12. DOI: 10.1111/are.14984

Claude, E.B. And Gross, A., (2014). Use of Probiotics for improving soil and water quality in aquaculture ponds. Research gate, 101-105.

FAO., (2020). The State of World Fisheries and Aquaculture (SOFIA); Fisheries and Aquaculture Department, Rome.

Fontenot, Q., Bonvillain, C., Kilgen, M. And Boopathy, R., (2007). Effects of temperature, salinity, and carbon: nitrogen ratio on sequencing batch reactor treating shrimp aquaculture wastewater. Biores. Tech., 98: 1700–1703.

Hari, B., Madhusoodana, K.B., Johny, T., Varghese, J.W., Schrama and Verdegem, M.C.J., (2006). The effect of carbohydrate addition on water quality and the nitrogen budget in extensive shrimp culture systems. Elsevier Aquaculture, 252: 248-263.

Huang, K., Wang, W. And Lu, J., (2003). Protein requirements in compounded diets for Penaeus vannamei juveniles China. J. Fish Sci., 10: 318-324.

Islam, M.L., Alam, M.J., Rheman, S., Ahmed, S.U. And Mazid, M.A., (2004). Water quality, nutrient dynamics and sediment profile in shrimp farms of the Sundarbans mangrove forest, Bangladesh. Indian J. Mar. Sci., 33(2): 170-176.

Menz, A. And Blake, B. F., (1980). Experiments on the growth of Penaeus vannamei Boone. J. Exp. Mar. Biol. Ecol., 48: 99–111.

MPEDA (Marine Products Export Development Authority), (2021). Annual report, Ministry of Commerce & Industry. Government of India.

Muralidharan, C.M., (2019). Need for ecosystem approach to brackish water shrimp farming - A case from Andhra Pradesh, India. book of abstracts, ICAR-Central Institute of Brackishwater Aquaculture and Society of Coastal Aquaculture and Fisheries, 49-52.

Muthusamy, R., Pramod, K.P., Radhakrishnapillai, A., Alagarsamy, V., Vivekanand, B. And Chandra, S.P., (2016). Effect of different biofloc system on water quality, biofloc composition and growth performance in Litopenaeus vannamei (Boone,1931). Aquaculture Research, 47: 3432–3444.

NFDB (National Fisheries Development Board)., (2020). Annual report, Department of Fisheries, Ministry of Fisheries, Animal Husbandry and Dairying.

Panigrahia, A., Saranyaa, C., Sundarama, M., Vinoth, K.S.R., Rasmi, R.D., Satish, K.R., Rajesha, P. And Otta, S.K., (2018). Carbon: Nitrogen (C:N) ratio level variation influences microbial community of the system and growth as well as immunity of shrimp (Litopenaeus vannamei) in biofloc based culture system. Fish and Shellfish Immunology, 81: 329–337.

PERez, F.I. And Kensley, B., (1997). Penaeoid and Sergestoid shrimps and prawns of the world. Key and diagnoses for the families and genera. Memoires du Museum national Histoire naturelle., 175:1-233.

Piedrahita, R.H., (2003). Reducing the potential environmental impact of tank aquaculture effluents through intensification and recirculation. Aquaculture, 226: 35-44.

Rivera, V.G., Soto, L.A., Salgado, U.I.H. And Naranjo, E.J., (2008). Growth mortality and migratory pattern of white shrimp (Litopenaeus vannamei, Crustacea, Penaeidae) in the carretas- pereyra coastal lagoon system, Mexico. Revt. Biol. trop., 56: 523-33.

Schneider, O., Sereti, E.H.V., Eding and Verreth, J.A.J., (2005). Analysis of nutrient flows in integrated intensive aquaculture systems. Aquacul. Eng., 32: 379-401.

Thitamadee, S., Prachumwat, A., Srisala, J., Jaroenlak, P., Salachan, P.V. And Sritunyalucksana, K., (2016). Review of current disease threats for cultivated penaeid shrimp in Asia. Aquaculture, 452: 69–87.

Vijayan, K.K. And Kailasam, M., (2020). Emerging trends of brackishwater aquaculture in India - Opportunities, challenges and way forward. ICAR-Central Institute of Brackish water Aquaculture Chennai, 2: 6-14.

Wujie, X., Timothy, C.M. And Tzachi, M.S., (2016). Effects of C/N ratio on biofloc development, water quality and performance of Litopenaeus vannamei juveniles in a biofloc-based, high-density, zero-exchange, outdoor tank system. Aquaculture, 453: 169-175.

Zhu, C., Dong, S.L. and WANG, F., 2006. The interaction of salinity and Na/K ratio in seawater on growth, nutrient retention and food conversion of juvenile Litopenaeus vannamei. J. Shellfish Res., 25(1): 107-112.

15

Recirculatory Aquaculture System (RAS) and its Challenges to Fish Farmers

Panchakarla Sedyaaw[1]*, Suhas Wasave[2], Sangita Wasave[3] and Ayushi Pandey[4]

[1]Department of Fish Processing Technology, College of Fisheries Ratnagiri-415629, Maharashtra, India

[2]Department of Fisheries resources extension, Statistics and economics College of Fisheries, Ratnagiri-415629, Maharashtra, India

[3]Department of aquaculture, College of Fisheries, Ratnagiri-415629 Maharashtra India

[4]Department of Fisheries Resource Management, College of Fisheries Ratnagiri-415629 Maharashtra, India

Abstract

Recirculatory Aquaculture Systems (RAS) represent a transformative approach to sustainable aquaculture, offering a myriad of opportunities while simultaneously presenting notable challenges. This abstract provides a concise overview of the key aspects shaping the landscape of RAS, focusing on its potential advantages and the obstacles faced by practitioners.RAS is positioned as a solution to address critical issues in traditional aquaculture by continuously filtering and recirculating water. This conservation-oriented design not only mitigates water usage but also contributes to sustainable practices, addressing environmental concerns and aligning with global efforts to enhance resource efficiency.The biosecurity advantages of RAS are notable, as the controlled environment allows for meticulous regulation of water quality parameters. This capability minimizes the risk of disease and promotes a healthier and more resilient aquaculture system. Moreover, precision control over environmental conditions within RAS fosters optimal growth conditions, resulting in increased production yields. This aspect is particularly significant in meeting the rising global demand for seafood, while ensuring the sustainability of aquaculture practices, RAS faces the challenges that hinder its widespread adoption. High initial investment costs constitute a significant barrier, necessitating strategic approaches to enhance economic viability and accessibility, especially for small-scale farmers.

Energy consumption and environmental impact concerns are also salient, prompting ongoing research into technologies that minimize ecological footprints, technical complexity, and the demand for skilled labor, further complicating the implementation of RAS. To address these challenges, the development of user-friendly interfaces and training programs has become essential, empowering aquaculturists to navigate RAS operations. RAS holds great promise for sustainable aquaculture, and concerted efforts are required to overcome these challenges and maximize its potential benefits on a global scale.

1. Introduction

Recirculatory Aquaculture Systems (RAS) have garnered attention as a sustainable solution to address the environmental and economic challenges associated with traditional aquaculture practices. Recent studies, such as those conducted by Martins *et al.* (2023) and Chen *et al.* (2024), have underscored the efficiency and resource conservation benefits of RAS. This section introduces the overarching opportunities and challenges in RAS, laying the foundation for a detailed examination of the latest developments in the field, opportunity presented by RAS lie in its capacity for water conservation. Martins *et al.* (2023) highlight the continuous filtration and recirculation of water within the system, significantly reducing overall water usage. This not only addresses environmental concerns but also aligns with the broader goals of sustainable aquaculture and provides a unique advantage in biosecurity, as emphasized by Li *et al.* (2022). The ability to meticulously control water quality parameters minimizes the risk of diseases and contributes to a healthier and more resilient aquaculture system. This enhanced biosecurity is pivotal in mitigating the impact of disease outbreaks on aquaculture operations. Precision control over environmental factors within the RAS facilitates optimal growth conditions, resulting in higher production yields. Anderson and Smith (2023) delved into the advantages of this controlled environment, positioning the RAS as a key driver for meeting the growing global demand for sustainable seafood production.

Aquaculture is the culture of aquatic animals and plants under controlled or semi-controlled conditions, that is underwater agriculture. It is the world's escalating food-producing sector and possesses an appeal that makes it one of the most efficient sectors to supply high-quality animal protein with low environmental impact (Tidwell and Bright, 2018). Aquaculture in ponds requires an outsized number of water resources and acreage and produces polluted effluent, which affects the environment in the future (Lin *et al.* 2003). Consistent with Losordo, Masser, and Rakocy for 1 acre of water area, approximately 1 million gallons of water is required to fill the pond and to compensate for evaporation and seepage during the year. This can be diminished by using the RAS within the fish-farming business. RAS is one of the most popular technologies worldwide. It is a technology used to farm aquatic organisms by reusing water in production based on the use of mechanical and biological filters (Bregnballe, 2015).

The system mostly uses tanks for production and thus requires a relatively small production area (Losordo *et al.* 1998). The recirculating system filters and cleans the water used for fish culture and recycles it back to the fish culture tanks (Helfrich and Libey, 1990). A high volume of water in the production system is circulated through mechanical or biological filters to be reused in tanks (Almeida *et al.* 2019). For the rapid growth of the fish in the system, fish should be fed high-protein pelleted diets with CP ranging from 1.5 to 15 percent of their body weight per day, according to the size and species of cultured fish. In this system, almost all organic matter produced within the system from uneaten feed/diets, dead bodies, and excreta of fish is not thrown with the effluents but is mineralized by heterotrophic bacteria by both the filter materials and rearing water (Losordo *et al.* 1998; Sugita *et al.* 2005). The system can be designed in a way that can raise large quantities of fish in a relatively small volume of water where water is treated to remove toxic waste products and then reusing the water (Farghally *et al.* 2014).

Fish are a valuable source of nutrients (e.g., proteins, essential fatty acids, and trace elements) for human nutrition and play an essential role in global food security (Tacon and Metian, 2013). According to the Food and Agriculture Organization (FAO), fish accounted for 17% of animal proteins and 7% of all proteins in the global population in 2017 (FAO, 2020). Aquaculture has been the main supplier of fish for human consumption since 2014, accounting for 52% of the total in 2018 (FAO, 2020). China is the world's largest aquaculture producer: in 2020, it produced 49.90 million tonnes of fish, accounting for 57.03% of the global total (FAO 2022). However, the development of Chinese aquaculture has been confronted with many problems owing to the excessive use of traditional culture systems (e.g., ponds and cages), such as disease outbreaks, environmental pollution, and food safety concerns (Cao *et al.* 2007; Liu *et al.* 2017). These problems have severely restricted the sustainable development of the Chinese aquaculture industry. There is an urgent need to improve this situation by developing culture systems in which fish production can be highly controllable and environmentally friendly, such as in industrialized aquaculture (Lei *et al.* 2014). Industrialized aquaculture is an important future trend for aquaculture development in China owing to its advantages in saving water and land resources and promoting higher productivity and sustainability. However, little information on its current status is currently available. This paper reviews the working design, challenges, and advantages of RAS. Table 1 lists the major design parameters of the RAS.

Table 1. Major design parameters for RAS (Source: Francis Murray, John Bostock (University of Stirling) and David Fletcher (RAS Aquaculture Research Ltd, 2014) (Source: Nilav *et al.*, 2020)

Parameter	**Comments**
Salinity	This will depend on the requirements of the species, but marine systems have inherently more complex water chemistry and less efficient biofiltration.
Biomass & feed rate	It provides the information about the variation in biomass and the quantity of feed introduced to the system each day is generally the most important factor for system sizing.
Stock density	This is highly dependent on species selected, size range and water quality parameters, tank dimensions and perhaps water flow dynamics. Higher stocking densities generally imply more efficient utilisation of tank volume and overall facilities.
Production plan	The use of multiple batches involving staggered stocking and harvesting schedules is normal in RAS to optimise use of resources and maintain reasonably stable biomass
Water flow rates	These may be calculated in relation to biomass has been stocked as to provide a consistent supply of water per minute per kg or stock. Consideration of water velocities in relation to body length can be a useful design parameter.
Feed system	This will be specified based on volumes and feed rates required.
Biosecurity	A risk assessment needs to be carried out that considers factors such as species, potential pathogens, disease susceptibility, location and potential routes of infection. This will lead to decisions on disinfection and other biosecurity measures.
Water quality targets	Typical parameters include suspended solids, dissolved oxygen and carbon dioxide, ammonia, nitrite and nitrate, pH, alkalinity, salinity and temperature need to be set at the design stage to help define performance requirements for treatment equipment.
Monitoring & control	Computerised control systems can both help to reduce labour requirements and improve response to out of range conditions. Requirements for system monitoring will be based on design the criteria and water quality targets set, together with a risk assessment of potential points of system failure.
Waste treatment and disposal	The major waste stream from RAS is organic solids which frequently need dewatering and other treatment prior to disposal.

2. Brief history of development

Although it is still considered a recent innovation in its infancy, the basic technology of RAS has existed for over 65 years, with the first pioneering RAS research activity being conducted in Japan in the 1950s (Murray *et al.*,

2014). According to Espinal and Matuli (2019), RAS technology, including aquaponics, has developed over the past 40 years. In the 1970s, a German program demonstrated the feasibility of intensive carp

production in RAS, and subsequently, the Danish Aquaculture Institute undertook an innovative effort to develop further technical aspects of the RAS (Goldman, 2016). The idea for commercial fish production in RAS was first fostered in Denmark in the mid-1970s, and the first commercial RAS was then built in 1980 (Warrer-Hansen, 2015). The Danish efforts supported the development of one of initial commercial RAS industries, and specifically

for the production of the European eel (Anguilla Anguilla) (Goldman, 2016). This work inspired the subsequent development and uptake of RAS in other European countries in the late 1980s and the 1990s (Martins *et al.*, 2010). Over the last 25 to 35 years, significant and growing experience in designing, building, and operating RAS, particularly in Nordic countries, has been reported (Dalsgaard *et al.*, 2013). The initial success of the RAS-based European eel industry also inspired the development of RAS in North America (Goldman, 2016). In China, the marine RAS was initiated in the 1980s, and since then, China has made considerable progress in RAS (Ying *et al.*, 2015). Since the 2000s, RAS has been developed in Europe, North America, Australia, and other aquaculture-producing countries (Espinal and Matulic, 2019). The key milestones and steps in the development of the RAS are summarized in Table 2.

Table 2. The historical development of RAS in the world. (Source: Ahmed *et al.*, 2021)

Period	Development
1950s	Research on RAS first conducted in Japan
1970s	Roots of modern RAS through German program that demonstrated Goldman (2016) intensive carp production.
Mid - 1970s	Idea for RAS as commercial fish production was first fostered in Denmark
1980	First commercial RAS of European eel production was built in Denmark
Early 1980s	RAS design and technology used for catfish and eel production in the Netherlands
Early 1980s	Started research and development on RAS in North America
1980s	Marine RAS begun in China
1980s-1990s	Development of RAS in other European countries
2000-2020	Further development of RAS in Australia, Europe, North America, and other countries

Significant acceleration in the development of RAS technology has been observed over the last two decades (Espinal and Matulic, 2019), and RAS has become popular in recent years. RAS have been developed to grow fish with inadequate biophysical conditions, water scarcity, poor water quality, and unfavorable environments (Murray *et al.*, 2014). According to Malone (2013), RAS provides an alternative production method when environmental regulations, diseases, land availability, salinity, temperature, and water supply prevent more cost-effective alternatives. However, other factors have also stimulated the development and implementation of RAS. For example, RAS are increasingly being used for Mediterranean

marine fish and salmonid production cycles, particularly for juvenile stages, before being transferred into outdoor grow-out systems, such as cages or flow-through raceways (Bostock *et al.*, 2016; Clarke and Bostock, 2017). RAS can be used for broodstock and seedstock production, which can support cage and net-pen aquaculture (Malone, 2013). In Europe and North America, RAS has been developed as an alternative to cage culture of salmon (Murray *et al.*, 2014). RAS has also been developed to culture exotic fish species in order to avoid adverse effects on native species and biodiversity (Murray *et al.*, 2014).

3. Principle of RAS

The basic principles of RAS operation and production RAS are land-based indoor fish rearing facilities, where fish are stocked in tanks within a controlled environment, and where filtration is applied to purify water by removing metabolic wastes of stock before being recirculated into the system itself. Water purification was achieved through mechanical and/or biological filtration, sterilization, and oxygenation (Fig. 1). Different levels of sophistication and efficiency can be achieved; however, generally, all RAS have a high degree (>90%) of water reuse (Murray *et al.*, 2014). RAS provides opportunities to enhance waste management, reduce water usage, and reduce nutrient recycling (van Rijn, 2013).

Although RAS was initially developed and is ideally suited to produce freshwater as well as warm water fish species (e.g., channel catfish, striped bass, and tilapia), RAS are flexible and can be modified and adopted for operation with brackish and marine water as well as cold water species (Helfrich and Libey, 2000). Therefore, by decoupling fish production from the marine environment, the RAS may offer an alternative to traditional and netpen aquaculture (O'Shea *et al.*, 2019). RAS can also provide suitable environmental conditions for fish species that are sensitive to water quality (Zhang *et al.*, 2011). Although RAS has the potential to produce diverse seafood products, it is generally utilized to culture high-value fish (e.g., barramundi, catfish, eel, perch, prawn, salmon, seabass, seabream, shrimp, sturgeon, trout, and tuna) with high stocking densities and year-round production to offset high operational costs (Dalsgaard *et al.*, 2013; Murray *et al.*, 2014). Nevertheless, other fish species, including Arctic char, clarias, halibut, pangasius, tilapia, and turbot, are also commonly produced in RAS (Badiola *et al.*, 2018).

The selection of fish species can be "market-driven because of a high return on investment to keep the RAS profitable. The choice of fish also depends on the fast-growing and hardy fish in the RAS (Badiola *et al.*, 2018). RAS can be categorized into five types: (1) hatchery and growout, (2) breeding, (3) long-term holding, (4) short-term holding, and (5) display (Yanong 2012). Moreover, RAS can be incorporated into an "integrated agriculture-aquaculture" system, which is known as aquaponics (Martins *et al.*, 2010). Aquaponics is considered a particular type of RAS, where vegetable plants are included in fish (i.e., aquaculture and hydroponics) to provide water filtration and crop diversification (Goddek *et al.*, 2019; Rakocy *et al.*, 2006). RAS has greater control over production outcomes, and the productivity of RAS depends on the culture species, stocking densities, feeding rate, duration of the production cycle, and other management aspects. According to the available scientific literature, the stocking densities of RAS range from 70 to 120 kg/m3 with feed conversion ratio (FCR) values from 0.8 1.1. RAS can be of various sizes, including small, medium, and large (Helfrich and Libey, 2000), with a large-scale RAS typically capable of producing 400 to 500 stocking densities, and total production values are currently reported by some commercial producers. According to Bregnballe (2015), RAS is a highly productive intensive farming system, which generates vast quantities of fish (500 ton/ha/year) in a comparatively small volume of water. Because of higher production, RAS are often referred to "hyper" or "super" intensive farming (O'Shea *et al.*, 2019).

4. Design of RAS

The basic principle behind the RAS is to recirculate water through a flow-through fish farm by diverting the water supply through ponds or tanks. However, recirculation implies the treatment of some or all of the discharged water and returning it to the fish rearing system. In an RAS, a key design parameter is the ratio of recycled water to wastewater (i.e., the percentage of recycled water in the fish tank inflow water). The main functional parts of an RAS include the following: 1. growing tank, 2. sump of the particulate removal device and 3. biofilter, 4. oxygen injection with U-tube aeration, and 5. The water circulation pump (Fig. 1). Depending on the type of fish species to be cultured, a thermostat system must be installed to maintain optimum water temperature. Ozone and UV sterilization are used to reduce organic and bacterial loads.

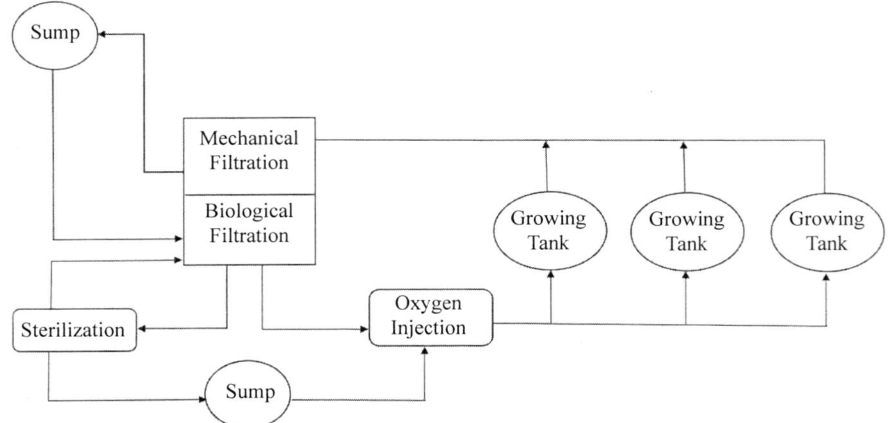

5. Components of RAS

5.1 Tank

Tank properties suitable for RAS should have a self-cleaning effect, low retention time of particles, oxygen control and regulation, and space utilization. The fish rearing tank should provide an environment that meets the needs of the fish, both in terms of water quality and tank design. The factors to be considered when choosing the right tank design are size and shape, water depth, self-cleaning ability, etc., which can have a considerable impact on the performance of the reared species and stocking density (Bregnballe, 2015). Brood stocks are commonly stocked at low density, whereas fingerlings, baitfish, and ornamental fish have moderate density, but grow-out fish can be highly dense (Malone, 2013). Accordingly, the circular tanks used in the system provide uniformity of the culture environment, allow a range of rational velocities that optimize fish health and conditions, and allow easy removal of settable solids (Farghally *et al.*, 2014).

The size of fish tanks is based on the density of the fish. There are three common tank shapes: circular, rectangular, and raceway. The dominance of circular tanks in the RAS industry stems from their inherent structural and hydrodynamic characteristics. The walls of a circular tank are maintained under tension by water pressure and are self-supporting. These properties allow circular tanks to be constructed from relatively thin polyethylene plastic or sturdier fiber glass materials. The hydrodynamics of a circular tank help in the rapid removal of suspended solids, which makes it more efficient than other tanks. A circular tank with a center drain was naturally suitable for solid removal. The rectangular tank is prone to poor solid movement, but it is approximately 20 percent more efficient in floor space utilization than the others. Raceway tanks would appear to be a perfect compromise between circular and rectangular tanks. A third wall is centered along the tank length to facilitate controlled circulation of water, which is highly effective for the movement of solids, but adds cost (Malone, 2013).

5.2 Circulation component

The fish tanks and filtration components were connected to circulation loops. Recirculation flow rates can be 5–10 gallons per minute per pound of daily feed ration provided to cultured fish. This varies according to the system's strategy. The major source of RAS energy consumption is the water pump or air blower that drives the circulation loop. If a circulation system failure occurs, it leads to rapid deterioration in the RAS tank water quality. Three common types of pumping systems were used: centrifugal, axial-flow, and airlift pumps. In most RAS applications, a centrifugal pump with high flow and low lift capacity is preferred to minimize energy consumption. Axial flow pumps are used in commercial larger-scale RAS because they have better pumping efficiencies than centrifugal pumps under low lift conditions (<10 feet or 3 m). Airlift pumps can move large volumes of water at extremely low lifts. The air injected to move the water also aerates and degasifies circulating water (Malone, 2013).

5.2.1 Physical Filters

Settable solids are the easiest to remove and should be removed as rapidly as possible from the tank through sedimentation tanks (clarifier), mechanical filters (granular or screen), or swirl separators (Losordo, 1998). Physical filters filter out relatively large coarse impurities such as feces, mucus, and leftover feed, despite the use of a chemical filter that filters out particles that cannot be processed by physical filters, that is, charcoal. Mechanical filtration of outlet water has been proven to be a practical solution for the removal of organic waste, but almost all recirculated farms filter outlet water using a microscreen fitted with a filter cloth typically 40–100 μm (Bregnballe, 2015). Using a microscreen showed significant improvement in water quality compared to the control groups in an experiment conducted by Fernandes *et al.* (2015). However, water quality using a 20 μm did not improve significantly as compared to the 100 μm mesh size. This might be due to prolonged operation under constant conditions, which weakens the mesh size to 20 μm and leads to cake formation, disrupting efficient water removal. In some systems, different types of filters, such as screened sedimentation, up-flow sand, and plastic bead filters,can be combined to accomplish mechanical filtration (Al-Hafedh *et al.*, 2003).

Mechanical filtration of the outlet water from the fish tanks has proven to be the only practical solutions to remove organic waste products. Today, almost all recirculated fish farms filter the outlet water from tanks in a microscreen fitted with a filter cloth of typically 40–100 μm. The drum filter is by far the most commonly used type of microscreen, and the design ensures the gentle removal of particles. The filtered water enters the drum and is filtered through the filter elements of the drum. The difference in the water levels inside and outside the drum is the driving force for filtration. The solids are trapped on the filter elements and lifted to the backwash area by rotating the drum. The rejected organic material was washed out of the filter elements and accumulated in a sludge tray. Reduction in organic load of the biofilter. Making the water clearer as organic particles were removed from the water. (Bregnballe, 2015)

5.2.2 Biological Filters

Biological filters are the processors of nitrogen compounds in water, such as bio-balls and bio-rings, and are used to maintain acceptable water parameters for larvae and juveniles by decreasing ammonia concentrations (Tanjung *et al.*, 2019; Pedreira *et al.*, 2016). It is also used to remove dissolved organics (Malone, 2013). Biofilters are classified as suspended or fixed. Fixed-film reactors are more stable than suspended growth systems (Malone and Pfeiffer 2006). Different types of plastic biofilter media with different configurations are available, that is, plastic rolls, PVC pipes, and scrub pads, which have different organic waste removal efficiencies from the effluents in the RAS. Plastic rolls tended to have a higher TAN (Total Ammonia Nitrogen) reduction rate than PVC pipes and scrub pads (Al-Hafedh *et al.*, 2003). However, all filter media are known to sustain the water quality within a suitable range for fish growth. Plastic rolls have many open areas for easy passage of water, which makes oxygen dissolution efficient, and the bacteria are never under oxygen stress (Al-Hafedh *et al.*, 2003). The efficiency of a biofilter depends on the water temperature and pH of the system (Bregnballe, 2015). Bioballs have a larger surface area, so the decomposition of ammonia is better as a greater surface area facilitates the surface for bacterial growth, which decomposes ammonia (Tanjung *et al.*, 2019).

Biological filtration is the core component of RAS. Nitrogen, in the form of free ammonia (NH3), is toxic to fish and must be transformed into a biofilter to harmless nitrate. Heterotrophic bacteria oxidize organic matter by consuming oxygen and producing carbon dioxide, ammonia, and sludge. Nitrifying bacteria convert ammonia into nitrite and finally into nitrate. The efficiency of biofiltration depends primarily on the water temperature and the pH level of the system. To achieve a satisfactory nitrification rate, water temperatures should be maintained within 10–35 °C (most favorable around 30 °C) and pH levels between 7 and 8 (Bregnballe, 2015).

6. Foam Fractionator

Some suspended solids that cannot settle at the bottom but exist in the water column may obstruct the gill function of the fish and limit their growth. Fine solids increase the oxygen demand in the system and cause gill irritation, which can be removed using a foam fractionator, which is also referred to as a protein skimmer, and it also helps to control the foaming agents that can accumulate in systems with long water reuse (Losordo *et al.*, 1998; Malone, 2013). Biological filters are the processors of nitrogen compounds in water, such as bio-balls and bio-rings, and are used to maintain acceptable water parameters for larvae and juveniles by decreasing ammonia concentrations (Tanjung *et al.*, 2019). It is also used to remove dissolved organics (Malone, 2013). Biofilters are classified as suspended or fixed. Fixed-film reactors are more stable than suspended growth systems (Malone and Pfeiffer 2006). Different types of plastic biofilter media with different configurations are available, that is, plastic rolls, PVC pipes, and scrub pads, which have different organic waste removal efficiencies from the effluents in the RAS. Plastic rolls tended to have a higher TAN (Total Ammonia Nitrogen) reduction rate than PVC pipes and scrub pads (Al-Hafedh *et al.*, 2003). However, all filter media are known to sustain the water quality within a suitable range for fish growth. Plastic rolls have many open areas for easy passage of water, which makes oxygen dissolution efficient, and the bacteria are never under oxygen stress (Al-Hafedh *et al.*, 2003). The efficiency of a biofilter depends on the water temperature and pH of the system (Bregnballe, 2015). Bioballs have a larger surface area, so the decomposition of ammonia is better as a greater surface area facilitates the surface for bacterial growth, which decomposes ammonia (Tanjung *et al.*, 2019).

According to an experiment conducted, the greater larval performance of Nile Tilapia and water quality were provided by the use of quartz gravel and porcelain as biofilters (Pedreira *et al.*, 2016). In an experiment conducted by (Singh and Wheaton, 1999), trickling biofilters performed better as compared to those in systems with bead filters). The advantages of trickling filters compared to other filters are that the stability of the process is high due to constant high oxygen levels, degassing removes CO^2, simple design, construction, operation, and management, and water cooling in summer (Eding *et al.*, 2006). Bacteria that grow in the recirculation system with filter media convert harmful compounds (Ammonia and Nitrite) into compounds that are not harmful to fish. Nitrate (Tanjung *et al.*, 2019). Most systems require a nitrification process to remove ammonia and nitrite from water. A group of researchers demonstrated that the addition of nitrifying bacteria products in the system improves the nitrification efficacy of the system, that is, reduced amounts of ammonia and nitrite and increased nitrate levels (Kuhn *et al.*, 2010). Biological treatment can be explained as the process by which bacteria are used to convert the dissolved waste present in the culture tanks to cell mass and other stable products (Golz, 1995). The control of the nitrification process in the RAS by a better understanding of nitrifying populations is important and can increase production volume, reduce discharged water, and enhance the profitability of system owners and managers (Auffret *et al.*, 2013).

6.1 Nitrification results

NH4 (ammonium) + 1.5 O2 $\rightarrow NO_2$ (nitrite) + H_2O + 2H+ + 2e

NO_2 (nitrite) + 0.5 $O_2 \rightarrow NO_3$ (nitrate) + e

$NH_4 + 2\ O_2 \leftrightarrow NO_3 + H_2O + 2H+$

The nitrification capability of a system depends on the biofilm that is formed. The biofilm formed on the biofilters is seen as a bacterial attachment site that encounters a variety of flows and qualities while maintaining their innate ability to process waste (Malone and Pfeiffer, 2006). The major benefit of biofilm formation is that biofilms provide protection from the effects of an adverse environment and host immune defenses (King, 2001). Biofilms can be found in different RAS materials, including fiberglass, plastic, PVC, glass, stainless steel, rubber, aluminum, foam, and cement (King, 2001). The biofilter, in which the media remains stationary and untouched, remains largely undisturbed because no filter backwash is required; however, in the bead filter, the media is frequently backwashed to remove entrapped solids, causing significant damage to the biofilm, which indicates a significant delay in the re-establishment of Nitrobacter in bead filter media (Singh *et al.*, 1999).

6.2 Aerators

The oxygen concentration within the tank is also important. The required concentration of oxygen can be supplied to the system through continuous aeration, either with atmospheric oxygen (air) or pure gaseous oxygen using aerators and air diffusor systems (Losordo *et al.*, 1998)

6.3 Pump

A group of researchers described that to achieve low cost, only one pump is used in their overall design to connect the culture tank with a reserved tank (Lee *et al.*, 2013). Direct injection of oxygen can also be performed by diffusers in the tank (Bregnballe, 2015). This leads to sufficient oxygen in the tank to maintain suitable growth conditions for fish.

6.4 Ozone Disinfection

The oxygen concentration within the tank is also important. The required concentration of oxygen can be supplied to the system through continuous aeration, either with atmospheric oxygen (air) or pure gaseous oxygen, using aerators and air diffusor systems (Losordo *et al.*, 1998).

A group of researchers described that to achieve low cost, only one pump is used in their overall design to connect the culture tank with a reserved tank (Lee *et al.*, 2013). Direct injection of oxygen can also be performed by diffusers in the tank (Bregnballe, 2015). This leads to sufficient oxygen in the tank to maintain suitable growth conditions for fish.

After 3 to 4 h of feeding the fish, the concentrations of ammonia, dissolved organics, and other waste products reached a maximum, which is a potential time for ozone application. Depending on the feeding frequency, the waste level differs; thus, a series of ozone treatments can be introduced into the system (Goncalves and Gagnon, 2011). The amount of ozone required for the treatment of water in an RAS is usually calculated using daily feed rate data. Ozone efficiently destroys bacteria, viruses, fungi, algae, and protozoa by unruly functioning as a cell membrane, entering the cell, and terminating the nuclear chemistry of the cell (Lawson, 1995). The use of ozone in freshwater systems can be beneficial, but caution should be exercised when using it in seawater or saltwater systems. Ozone is very reactive to bromide and chloride ions in saltwater to form toxic hypobromite and hypochlorite ions. Therefore, residuals after the ozone treatment of saltwater should be removed or removed before the water is reused in the systems (32 and 31). In addition, ozone can cause depletion of trace elements in saltwater, especially manganese and calcium (Lawson, 1995).

7. Water quality

Fish are raised in an environment where they are exposed to their waste, which can quickly become toxic to them (Golz, 1995). All production systems must provide an appropriate environment to market fish growth with optimal values of critical water quality parameters such as dissolved oxygen, un-ionized ammonia-nitrogen, nitrite-nitrogen, and CO_2 in water (Losordo *et al.*, 1998). The reduced amount of makeup water, that is, new water entering the system, causes system flush to reduce, and water quality within the system is consequently degraded (Good *et al.*, 2009). Water quality issues are difficult to assess because they are caused by various factors, such as poor approach to the overall system and production quantities, equipment failure, or poor maintenance of the system (Badiola *et al.*, 2012).

Water quality reflects the overall capability of culture water to provide optimal growth conditions for the species of interest and that's why it is very important. The feeding behaviour of fish is strongly influenced by environmental conditions such as water temperature, DO, TAN, and NO_2–N (Pang *et al.*, 2011). RAS provides better environmental conditions year-round, contributes to the health of the fish, and minimizes the FCR, thus improving feeding efficiency (D'Orbcastel *et al.*, 2009). The RAS showed significant effects on various water quality parameters such as dissolved oxygen (DO), pH, Total Suspended Solids (TSS), Chemical Oxygen Demand (COD), Total Ammonium Nitrogen (TAN), Nitrite and Nitrate levels in the fish tank, as described below. Table 3,4 and 5 provide water quality standards for fish production using RAS technology.

Table 3. Water quality standards for fish production

Parameters	**Acceptable Concentration**
DO	>5 mg/L
PH	6.5-8.5
Temperature	>20°C for warm water species 15-20°C for cool water species
COD	20-30 mg/L
TSS	<80 mg/L
Nitrite (NO_2)	<0.02
Nitrate (NO_3)	0-100
TAN	0-0.2

(*Source:* Bhatnagar *et al.,* 2013)

Table 4. Guidelines for recommended water quality requirements of recirculating system

Component	**Recommended value or range**
Temperature	The optimum range for species cultured-less than 5°F as rapid change
Dissolved oxygen	60% or more of saturation usually 5ppm or more to warm water fish- >2 ppm in biofilter effluent
Carbon dioxide	Less than 20 ppm
pH	7.0 to 8.0
Total alkalinity	50 ppm or more
Total hardness	50 ppm or more
Un-ionized ammonia	Less than 0.05 ppm
Nitrite	Less than 0.005 ppm

(*Source:* Balami 2019)

Table 5. Key features of fish production in RAS (values are indicative and from published scientific papers; higher values are reported by some commercial operations)

Feature	**Information**
Stocking density (kg/m3)	70-120
Duration of production cycle- salmon (month)	9-14
Feed conversation ratio (FCR)	0.8-1.1
Production (ton/year)	400-500

(*Source:* A hmed *et al.,* 2021)

7.1 Advantages

The Recirculating Aquaculture System partially reuses water after proper treatment, thereby reducing water usage and improving effluent quality (Ramirez-Godinez *et al.*, 2013). This system provides potential advantages over a pond or cage culture, such as flexible site selection, less water usage, lower volume of effluent that provides better environmental management, provides a higher intensity of production, and has better environmental control (Goncalves and Gagnon, 2011). One of the abilities of RAS is to regulate temperature, which is advantageous as it accelerates the development of various reared fish and also avoids the seasonal prevalence of fish (Mongirdas *et al.*, 2017). Despite being water-conserving, RAS offers many advantages, such as less disease occurrence, improved feed conversion, a shorter production cycle due to a controlled environment, and consistent product quality (Singh *et al.*, 1999). These systems are uniquely engineered ecosystems that help minimize environmental perturbations by reducing nutrient pollution discharge (Bartelme *et al.*, 2017).

It is also an appropriate system that provides minimal effluent discharge, efficient water reuse, and optimal water conservation (Rakocy *et al.*, 2006). Grouper (Epinephelus coioides) juveniles with a stocking density of 25 fish/L showed the highest growth compared to 15 and 20 fish/L when reared for 70 days in an experiment conducted without affecting the survival rate of fish in all treatments, proving that RAS can be used to increase production (Agus *et al.*, 2014). The study also indicated that the greater effects on fish growth performance, water quality, and feed utilization were due to RAS and stocking density. Farmers can also benefit from combining RAS with plant culture, that is, aquaponics. The plant grows rapidly using the dissolved nutrients in fish cultured water, which is directly excreted by fish or generated by the microbial breakdown of fish waste (Rakocy *et al.*, 2006). Farmers can sell plants grown in aquaponics and earn extra income. The vegetable varieties that can be grown include tomato, lettuce, cucumber, squash, and Chinese cabbage.

7.2 Challenges

RAS are capital-intensive operations that require high funding for equipment, infrastructure, influent and effluent, treatment systems, engineering, construction and management. In this system, there is continuous circulation of the same water, which challenges the prevention and treatment of diseases. Pathogens spread throughout the system, and the addition of chemicals and antibiotics can disrupt the microbiome of biofilters (Almeida *et al.*, 2019). Failure of the biofilter may result in varying levels of ammonia or nitrite, both of which are toxic to fish and may result in health issues, suppressed growth, and mortalities of cultured aquatic animals. Low water exchange RAS have a greater chance of infections in aquatic animals than high water exchange systems, such as flow-through tanks (Good *et al.*, 2009). Although RAS has many advantages, it poses the potential risk of latent disease and public health risks. As the water is reused, pathogens introduced into the system could remain incorporated into the biofilm, feeding to recurring exposure of fish to pathogens and the presence of asymptomatic carriers. The most significant human pathogens are *Bacillus cereus, Shigella species,* and *Vibrio species,* all of which are responsible for gastrointestinal diseases (King, 2001).

7.2.1 Biosecurity and disease occurrence in RAS system

Biosecurity implies any company policy and procedures used on a farm that reduce the risk of pathogen introduction or spread through the facility if introduced. One of the primary advantages of RAS technology is that it provides farmers with the opportunity to reduce disease outbreaks and eliminate some diseases. However, while RAS can create optimum conditions for fish culture, inferior designs may inadvertently provide favorable conditions for disease outbreaks or reproduction of opportunistic pathogens (Delabbio *et al.*, 2004; Timmons *et al.*, 2010).

Once the pathogens enter the RAS, their potential impact on the stock can be influenced by the quality of the system design, but equally importantly, by the knowledge and experience of the RAS manager. D'Orbcastel *et al.*, (2009) evaluated RAS trout farms and noticed that the sedimentation system showed good but highly variable removal efficiency (60%), such that the remaining suspended solids were circulated and degraded in the system. This results in sedimentation areas in other regions of the RAS and general water quality degradation, and the efficiency of the biofilter was also variable due to the lack of temperature control. Excessive suspended solids can disturb the N_2 cycle and directly lead to nitrite toxicity and mass mortality (Mirzoyan *et al.*, 2010). Accumulation of nutrients and dissolved organic materials originating from uneaten feed and fish feces can create a favorable environment for a diverse range of bacteria, protozoa, micrometazoa, dinoflagellates, and fungi, which can alter the water quality and subsequently the stock (Michaud *et al.*, 2014; Michaud *et al.*, 2006).

7.2.2 Parasites in RAS

Both low- and high-tech RAS farms may become infected with pathogens, irrespective of the level of control over water quality and biosecurity precautions. Even the most efficiently operated farms may eventually become contaminated with a range of monogenean, protozoan, and dinoflagellate parasites. According to the design of the RAS farm and the technology used, farms infected with parasites may still have the potential to infect recipient waters according to the manner or efficiency of farm effluent management. The most commonly occurring parasites in the RAS system are several ciliated protozoan species. *Trichodina spp., Apiosoma sp., Ambiphrya sp., Epistylis sp., Chilodonella piscicola,* and *Icthyobodo necator.* Other more complex parasites of trout include *Spironucleus salmonis (Diplomonadida), Gyrodactylus derjavinoides (monogenean platyhelminthes),* and the eye fluke *Displostomum spathaceum (digenean)*. In Danish and European marine RAS farms infested with *Luciella masanensis*, fish mortality increased dramatically despite treatment of water with peracetic acid and chloramine-T. In another brackish water RAS farm infected by *Pfiesteria shumwayae*, the water was treated with chloramine-T, which caused dinoflagellates to temporarily disappear from the water column, apparently forming temporary cysts. (Nilav *et al.*, 2020)

7.2.3 Harmful Algal Blooms (HABs) in RAS

Some of the HABs species are directly parasitic, while others can impact stock by releasing harmful toxins within the RAS or in the source waters. The broad chemical and structural diversity of algal toxins, coupled with differences in their intrinsic potency and susceptibility to biotransformation, account for many of the challenges associated with the detection of these compounds (Yanong, 2009). Technology capable of detecting HABs or toxic by-products would be a critical development for RAS with high biomass loads at elevated stocking densities. Moreover, the treatment of raw water prior to entering the RAS facility is a critical component of the RAS design in farms exposed to potential HAB blooms.

7.2.4 Microbial pathogens in RAS system

RAS units with or without poor disinfection facilities (UV and ozone) can be susceptible to potential microbial pathogen infestation, as bacteria, viruses, and fungi can cause severe threats to the RAS unit. The most commonly occurring bacteria that increase in number in recirculating systems include *Aeromonas spp., Vibrio spp., Mycobacterium spp., Streptococcus spp.,* and *Flavobacterium spp.* (Yanong, 2009). Interestingly, some viruses, such as IPNV, require dose rates that are 7.5 times higher than most bacteria (Yoshimizu *et al.*, 1986). The most effective precaution against important viral diseases is to ensure that eggs, larvae, or fry are obtained from specific pathogen-free facilities and to implement strict biosecurity measures. The use of up to 2 ppt salinity, in addition to UV or ozone disinfection, has been found to help minimize this problem. When chemical treatments are added to RAS water, the biofilters are often exposed to a high concentration of the chemical, which increases the risk of impairing the nitrifying microbial population and hence, reduces the performance of the biofilter (Schwartz *et al.*, 2000).

8. Future prospectus

Climate change, scarcity of freshwater, and sudden outbreaks of diseases pose a severe threat to the aquaculture industry. The world population is growing at a rate of 1.1 % per year, and is expected to reach 8.5 billons by the end of 2030. The FAO projected a 1.8 percent increase in the per capita consumption of fish food by 2030. This will put an additional burden on farm production, as marine capture fisheries are also expected to decline over time due to overfishing. To maintain the present growth of the aquaculture industry, an innovative approach is required to address these issues. The recirculating aquaculture system is a highly intensive culture technique that uses very little freshwater and facilitates full control over disease outbreaks and other external environmental factors affecting fish culture. With RAS, 30-50 times more fish production per unit area is possible compared to traditional fish farming with limited use of water.

Conclusion

In conclusion, the Recirculatory Aquaculture System (RAS) stands at the forefront of sustainable aquaculture, offering a range of opportunities that hold the promise of transforming the industry. The inherent water conservation features of RAS address a critical concern in traditional aquaculture, aligning with global initiatives to promote responsible water resource management. By continuously filtering and recirculating water within a closed system, RAS not only minimizes water usage but also mitigates the environmental impact associated with traditional open-water systems. The biosecurity advantages of RAS further underscore its significance in aquaculture. A controlled environment allows meticulous regulation of water quality parameters, reducing the risk of diseases and enhancing the overall health of aquatic organisms. This biosecurity aspect not only safeguards the well-being of the farmed species but also contributes to the sustainability and resilience of aquaculture operations. RAS's potential to provide increased production efficiency through precise environmental control is paramount. The ability to manage factors such as temperature, oxygen levels, and nutrient concentrations creates optimal growth conditions, contributing to higher production yields. As the global demand for seafood continues to rise, the efficiency gains offered by RAS position it as a key player in meeting these demands sustainably; however, the journey towards widespread adoption of RAS is not without its challenges. High initial investment costs remain a significant barrier, particularly for small-scale farmers, necessitating strategic approaches to enhance their economic viability and accessibility. Overcoming these financial hurdles is crucial for unlocking the full potential of RAS and ensuring that its benefits are realized across diverse aquaculture scales. Energy consumption and the associated environmental impact represent additional challenges that require ongoing research and innovation. Striking a balance between operational efficiency and environmental sustainability is essential for the continued success of RAS. Addressing the technical complexity and demand for skilled labor is another pivotal aspect. Developing user-friendly interfaces and comprehensive training programs will empower aquaculturists to navigate the intricacies of RAS operations, foster wider adoption, and ensure the long-term success of this transformative aquaculture methodology. In conclusion, while RAS presents a paradigm shift in aquaculture towards sustainability and efficiency, concerted efforts are needed to address challenges and create an enabling environment for its global acceptance and implementation. The future of aquaculture may be shaped by the continued evolution and integration of Recirculatory Aquaculture Systems into mainstream practices.

References

Agus, P.A.S., Nan, F.H., Lee, M.C., 2014. Effects of stocking density on growth and feed utilization of grouper (Epinephelus coioides) reared in recirculation and flow-through water system. African J. Agric. Res., 9,

Ahmed, Nesar & Turchini, Giovanni. (2021). Recirculating aquaculture systems (RAS): Environmental solution and climate change adaptation. Journal of Cleaner Production. 297. 1-14.

Al-Hafedh, Y.S., Alam, A., Alam, M.A., 2003. Performance of plastic biofilter media with different configuration in a water recirculation system for the culture of Nile tilapia (Oreochromis niloticus). Aquac. Eng., 29, Pp. 139–154.

Almeida, G.M.F., Mäkelä, K., Laanto, E., Pulkkinen, J., Vielma, J., Sundberg, L.R., 2019. The fate of bacteriophages in recirculating aquaculture systems (RAS)—towards developing phage therapy for RAS. Antibiotics., 8.

Anderson, J., & Smith, R. (2023). "Optimizing Production Yields in Controlled Environments: The Role of Recirculatory Aquaculture Systems." Sustainable Fisheries Journal, 22(1), 56-73.

Auffret, M., Yergeau, É., Pilote, A., Proulx, É., Proulx, D., Greer, C.W., Vandenberg, G., Villemur, R., 2013. Impact of water quality on the bacterial populations and off-flavours in recirculating aquaculture systems. FEMS Microbiol. Ecol., 84, Pp. 235–247.

Badiola, M., Basurko, O.C., Piedrahita, R., Hundley, P., Mendiola, D., 2018. Energy use in recirculating aquaculture systems (RAS): a review. Aquacult. Eng. 81, 57e70.

Badiola, M., Mendiola, D., Bostock, J., 2012. Recirculating Aquaculture Systems (RAS) analysis: Main issues on management and future challenges. Aquac. Eng., 51, Pp. 26–35.

Balami, S. (2019). Recirculation aquaculture systems: components, advantages, and drawbacks. Education, 2021.

Bartelme, R.P., McLellan, S.L., Newton, R.J., 2017. Freshwater recirculating aquaculture system operations drive biofilter bacterial community shifts around a stable nitrifying consortium of ammonia-oxidizing archaea and comammox Nitrospira. Front. Microbiol., 8.

Bhatnagar, A., & Devi, P. (2013). Water quality guidelines for the management of pond fish

Bostock, J., Lane, A., Hough, C., Yamamoto, K., 2016. An assessment of the economic contribution of EU aquaculture production and the influence of policies for its sustainable development. Aquacult. Int. 24, 699e733.

Bregnballe, J., 2015. A Guide to Recirculation Aquaculture. Food and Agriculture Organization of the United Nations and EUROFISH International Organization.

Cao L, Wang W, Yang Y, Yang C, Yuan Z, Xiong S, Diana J (2007).Environmental impact of aquaculture and countermeasures to pollution in China. Environ Sci Pollut R 14:452–462.

Chen, B., et al. (2024). "Efficiency Gains in Recirculatory Aquaculture Systems: A Comprehensive Review." Journal of Sustainable Aquaculture, 18(3), 245-261.

Clarke, R., Bostock, J., 2017. Regional Review on Status and Trends in Aquaculture Development in Europe e 2015. FAO Fisheries and Aquaculture Circular No. 1135/1, Rome. culture. International Journal Of Environmental Sciences, 3(6).

Dalsgaard, J., Lund, I., Thorarinsdottir, R., Drengstig, A., Arvonen, K., Pedersen, P.B., 2013. Farming different species in RAS in Nordic countries: current status and future perspectives. Aquacult. Eng. 53, 2e13.

Delabbio, J., B.R. Murphy, G.R. Johnson and S.L. McMullin. 2004. An assessment of biosecurity utilization in the recirculation sector of finfish aquaculture in the United States and Canada. Aquaculture, 242(1-4):165-179.

D'Orbcastel, E. R., Ruyet, J. P., Bayon, N. L., & Blancheton, J. (2009). Comparative growth and welfare in rainbow trout reared in recirculating and flow through rearing systems. Aquacultural Engineering, 40(2), 79-86.

Eding, E.H., Kamstra, A., Verreth, J.A.J., Huisman, E.A., Klapwijk, A., 2006. Design and operation of nitrifying trickling filters in recirculating aquaculture: A review. Aquac. Eng., 34, Pp. 234–260.

Espinal, C. A., & Matuli_c, D. (2019). Recirculating aquaculture technologies. In Aquaponics food production systems (pp. 35–76).

Espinal, C.A., Matuli_c, D., 2019. Recirculating aquaculture technologies. In: Goddek, S., Joyce, A., Kotzen, B., Burnell, G.M. (Eds.), Aquaponics Food Production Systems: Combined Aquaculture and Hydroponic Production Technologies for the Future. Springer Open, Switzerland, pp. 35e76.

FAO (2020) The state of world fisheries and aquaculture 2020. Sustainability in action. Food and Agriculture Organization of the United Nations, Rome.

FAO (2022) FishstatJ, a tool for fishery statistical analysis. global fishery and aquaculture production 1950–2020. Rome, Italy.

Farghally, H.M., Atia, D.M., El-madany, H.T., Fahmy, F.H., 2014. Control methodologies based on geothermal recirculating aquaculture system. Energy, 78, Pp. 826–833.

Fernandes, P., Pedersen, L.F., Pedersen, P.B., 2015. Microscreen effects on water quality in replicated recirculating aquaculture systems. Aquac. Eng., 65, 17–26.

Goddek, S., Joyce, A., Kotzen, B., Burnell, G.M. (Eds.), 2019. Aquaponics Food Production Systems: Combined Aquaculture and Hydroponic Production Technologies for the Future. Springer Open, Switzerland.

Goldman, J., 2016. So, you want to be a fish farmer? World Aquacult. 47 (2), 24e27.

Golz, W.J., 1995. Biological Treatment in Recirculating Aquaculture Systems 1. Aquaculture, Pp. 6–7.

Gonçalves, A.A., Gagnon, G.A., 2011. Ozone application in recirculating aquaculture system: An overview. Ozone Sci. Eng., 33, Pp. 345–367.

Good, C., Davidson, J., Welsh, C., Brazil, B., Snekvik, K., Summerfelt, S., 2009. The impact of water exchange rate on the health and performance of rainbow trout Oncorhynchus mykiss in water recirculation aquaculture systems. Aquaculture, 294, Pp. 80–85.

Helfrich, L.A., Libey, G., 1990. Fish Farming in Recirculating Aquaculture Systems (RAS). Dep. Fish. Wildl. Sci., 19.

Helfrich, L.A., Libey, G., 2000. Fish Farming in Recirculating Aquaculture Systems. RAS). Department of Fisheries and Wildlife Sciences, Virginia Tech, USA.

King, R.K., 2001. The presence of bacterial pathogens in recirculating aquaculture system biofilms and their response to various sanitizers. Fac. Virginia Polytech. Inst. State Univ., Pp. 131.

Kuhn, D.D., Drahos, D.D., Marsh, L., Flick, G.J., 2010. Evaluation of nitrifying bacteria product to improve nitrification efficacy in recirculating aquaculture systems. Aquac. Eng., 43, Pp. 78–82.

Lawson, T.B., 1995. Recirculating Aquaculture Systems. In: Chapman and Hall (eds.) Fundamentals of Aquaculture engineering, Pp. 192–247. Springer Science+Business Media Dordrecht, New York.

Lee, J.V., Loo, J., Chuah, Y.D., Tang, P., Tan, Y., Wong, C., 2013. The Design of a Culture Tank in an Automated Redirculating Aquaculture System. Int. J. Eng. Appl. Sci., 2, Pp. 67–77.

Lei J, Huang B, Liu B, Xu Z, Yan K, Zhai J (2014) Strategic research on the construction of high-end farming industry in China based on the concept of aquatic animal welfare. Strategic Study of CAE 16:14–20.

Li, C., et al. (2022). "Biosecurity Measures in Recirculatory Aquaculture: Case Studies and Best Practices." Aquaculture Research, 30(4), 421-438.

Lin, Y.F., Jing, S.R., Lee, D.Y., 2003. The potential use of constructed wetlands in a recirculating aquaculture system for shrimp culture. Environ. Pollut., 123, Pp. 107–113.

Liu X, Steele JC, Meng X-Z (2017) Usage, residue, and human health risk of antibiotics in Chinese aquaculture: a review. Environ Pollut 223:161–169.

Losordo, T.M., Masser, M.P., Rakocy, J., 1998. Recirculating aquaculture tank production systems: An overview of critical considerations. SRAC No, 451, Pp. 18–31.

Malone, R., 2013. Recirculating Aquaculture Tank Production Systems: A Review of Current Design Practice. Southern Regional Aquaculture Center, USA.

Malone, R.F., Pfeiffer, T.J., 2006. Rating fixed film nitrifying biofilters used in recirculating aquaculture systems. Aquac. Eng., 34, Pp. 389–402.

Martins, A., et al. (2023). "Advancements in Water Conservation Through Recirculatory Aquaculture Systems." Aquaculture Today, 45(2), 112-128.

Martins, C.I.M., Eding, E.H., Verdegem, M.C.J., Heinsbroek, L.T.N., Schneider, O., Blancheton, J.P., d'Orbcastel, E.R., Verreth, J.A.J., 2010. New developments in recirculating aquaculture systems in Europe: a perspective on environmental sustainability. Aquacult. Eng. 43, 83e93.

Michaud, L., J.P. Blancheton, V. Bruni and R. Piedrahita. 2006. Effect of particulate organic carbon on heterotrophic bacterial populations and nitrification efficiency in biological filters. Aquacultural Engineering, 34:224–233.

Mirzoyan, N., Y. Tal and A. Gross. 2010. Anaerobic digestion of sludge from intensive recirculating aquaculture systems. Aquaculture, 306(1-4): 1-6.

Mongirdas, V., Žibienė, G., Žibas, A., 2017. Waste and Its Characterization in Closed Recirculating Aquaculture Systems – a Review. J. Water Secur. 3.

Murray, F., J. Bostock and D. Fletcher. 2014. Review of recirculation aquaculture system technologies and their commercial application.

Nilav Aich, Suman Nama, Abhilipsa Biswal and Tapas Paul. 2020. A review on recirculating aquaculture systems: challenges and opportunities for sustainable aquaculture. Innovative Farming, 5(1): 017-024.

O'Shea, T., Jones, R., Markham, A., Norell, E., Scott, J., Theuerkauf, S.,Waters, T., 2019. Towards a Blue Revolution: Catalyzing Private Investment in Sustainable Aquaculture Production Systems. The Nature Conservancy and Encourage Capital, Virginia, USA.

Pang, X., Cao, Z., & Fu, S. (2011). The effects of temperature on metabolic interaction betweendigestion and locomotion in juveniles of three cyprinid fish (Carassius auratus, Cyprinus carpio and Spinibarbus sinensis). Comparative Biochemistry and Physiology Part A:Molecular & Integrative Physiology, 159(3), 253-260.

Pedreira, M.M., Tessitore, A.J. de A., Pires, A.V., Silva, M. de A., Schorer, M., 2016. Substrates for biofilter in recirculating system in Nile tilapia larviculture production. Rev. Bras. Saude e Prod. Anim., 17, Pp. 553–560.

Rakocy, J.E., Masser, M.P., Losordo, T.M., 2006. Recirculating aquaculture tank production systems: Aquaponics- integrating fish and plant culture. SRAC Publ. - South. Reg. Aquac. Cent., 2, Pp. 16.

Rakocy, J.E., Masser, M.P., Losordo, T.M., 2006. Recirculating Aquaculture Tank Production Systems: Aquaponics e Integrating Fish and Plant Culture. Southern Regional Aquaculture Center, USA.

Ramírez-Godínez, J., Beltrán-Hernández, R.I., Coronel-Olivares, C., Contreras-López, E., Quezada-Cruz, M., Vázquez-Rodríguez, G., 2013. Recirculating Systems for Pollution Prevention in Aquaculture Facilities. J. Water Resour. Prot. 05, Pp. 5–9.

Schwartz, M.F., G.L. Bullock, J.A. Hankins, S.T. Summerfelt and J.A. Mathias. 2000. Effects of Selected Chemotherapeutants on Nitrification in Fluidized-Sand Bioftlters for Coldwater Fish Production. International Journal of Recirculating Aquaculture, 1(1).

Singh, S., Ebeling, J., Wheaton, F., 1999. Water quality trials in four recirculating aquacultural system configurations. Aquac. Eng., 20, Pp. 75–84.

Sugita, H., Nakamura, H., Shimada, T., 2005. Microbial communities associated with filter materials in recirculating aquaculture systems of freshwater fish. Aquaculture, 243, Pp. 403–409.

Tacon AG, Metian M (2013) Fish matters: importance of aquatic foods in human nutrition and global food supply. Rev Fish Sci 21:22–38.

Tanjung, R.R.M., Zidni, I., Iskandar, Juniato, 2019. Effect of difference filter media on Recirculating Aquaculture System (RAS) on tilapia (Oreochromis niloticus) production performance. World Sci. News., 118, Pp. 194–208.

Tidwell, J.H., Bright, L.A., 2018. Freshwater aquaculture. In: Fath, B. D., (Eds.), Encyclopedia of Ecology. second ed., Elsevier Inc., Pp. 91-96.

Timmons, M.B., J.M. Ebeling, N.R.A. Center. 2010. Recirculating Aquaculture. Cayuga Aqua Ventures Ithaca, NY.

van Rijn, J., 2013. Waste treatment in recirculating aquaculture systems. Aquacult. Eng. 53, 49e56.

Warrer-Hansen, I., 2015. Potential for Land Based Salmon Grow-Out in Recirculating Aquaculture Systems (RAS) in Ireland. A Report to the Irish Salmon Growers' Association, Ireland.

Yanong, R.P. 2009. Fish health management considerations in recirculating aquaculture systems. 1: 1-9.

Yanong, R.P.E., 2012. Biosecurity in Aquaculture, Part 2: Recirculating Aquaculture Systems. Southern Regional Aquaculture Center, USA.

Ying, L., Baoliang, L., Ce, S., Guoxiang, S., 2015. Recirculating aquaculture systems in China e current application and prospects. Fish. Aquacult. J. 6 (3), 1000134.

Yoshimizu, M., H. Takizawa and T. Kimura. 1986. UV susceptibility of some fish pathogenic viruses. Fish Pathology, 21(1): 47-52.

Zhang, S.-Y., Li, G., Wu, H.-B., Liu, X.-G., Yao, Y.-H., Tao, L., Liu, H., 2011. An integrated recirculating aquaculture system (RAS) for land-based fish farming: the effects on water quality and fish production. Aquacult. Eng. 45, 93e102.

16

Histological Alterations in Gill, Kidney and Liver of Fish as a Diagnostic Tool for Ecotoxicological Research

Tiyasha Bhattacharya[1] and Avishek Bardhan[2*]

[1]College of Fisheries, Assam Agricultural University, Assam-781203, India

[2]Department of Aquatic Animal Health Management, Division of Fisheries Neotia University, Sarisha-743368, West Bengal, India

Abstract

Over the past few decades, water contamination has emerged as a significant global environmental concern affecting both freshwater and marine ecosystems. Anthropogenic pollutants, including non-essential metals (such as arsenic, cadmium, mercury, nickel, and lead), pesticides, persistent organic pollutants, and dioxins, are prevalent in water bodies. These pollutants not only accumulate in living organisms but also magnify through the trophic food chain, posing a serious threat to various species.

Owing to their persistence and resistance to degradation, these substances and thier toxicity are difficult to mitigate. To comprehend the detrimental effects on living organisms, particularly fish, scientists have employed biomarkers at different levels, ranging from cellular to population scales. At the tissue level, histological alterations in fish organs such as the gills, liver, and kidneys serve as valuable indicators of xenobiotic exposure. These histological changes offer crucial insights into the effects of contaminants on the health of fish. Consequently, they have become essential biomarkers in ecotoxicological research, aiding risk assessment and monitoring programs. This chapter provides a brief yet comprehensive review of scientific data, highlighting the significance of histological alterations in crucial fish organs as essential tools for understanding the ecological impact of water pollutants.

***Keywords**: Contamination; Fish; Biomarkers; Histopathology; Ecotoxicology*

1. Introduction

Human activities significantly lead to deterioration of ecological systems, affect individual organisms, and disrupt natural ecosystems. The rise in foreign compounds, such as heavy metals and pesticides, in aquatic ecosystems due to human activities has spurred extensive global research efforts (Wester *et al.,* 2002). The Union's Water Framework Directive (WFD) mandates monitoring programs to assess water body health. In environmental studies, understanding the impact of contaminants on aquatic life requires a multifaceted approach (Ullah and Zorriehzahra, 2015). Biomarkers that observe biological responses at various levels offer crucial insights into the presence and bioavailability of toxicants, complementing traditional methods (Yancheva *et al.,* 2016). These indicators, termed early warning signals, are vital for evaluating exposure, effect mechanisms, susceptibility to contaminants, and enhancing predictive models (Yancheva *et al.,* 2016). Integrating biomarkers into ecological studies provides a holistic understanding of environmental stressors and their implications in aquatic organisms.

2. Why is fish used as a model organism in ecotoxicological studies?

According to the Water Framework Directive (WFD), fish serves as a crucial indicator of river ecological status (Ullah and Zorriehzahra 2015). Their diverse sizes, ages, trophic levels, and sensitivity to various toxicants make them excellent indicators of water contamination (Wester *et al.,* 2002). Fish adapt their metabolic functions in response to environmental changes, which makes them valuable in toxicological research. Monitoring sentinel fish species is a widely accepted method for assessing toxicant accumulation and its impact on human health (Wester *et al.,* 2002). Fish, with their developed osmoregulatory, endocrine, nervous, and immune systems, are preferred to invertebrates in toxicological studies. Fish can absorb toxicants through both waterborne and dietary exposure, allowing the evaluation of contaminant transfer through the food chain. Key tissues for ecological, toxicological, and pathological studies in teleost fish include the gills, liver, and kidney because of their high metabolic activity and tendency to accumulate toxicants (Wester *et al.,* 2002).

2.1 Histological alterations as biomarkers

Histopathology, the microscopic study of cells and tissues, is pivotal for detecting histological abnormalities caused by xenobiotic effects in aquatic organisms (Ferreira *et al.,* 2010). These alterations, observed in specific organs, serve as sensitive biomarkers, offering early and nuanced insights into the effects of aquatic pollution, surpassing individual biochemical parameters

(Poleksić *et al.,* 2010). Histological analysis of fish tissues has long been a cornerstone of aquatic toxicology, providing valuable information alongside physicochemical analyses.

Histopathological changes in fish organs have been instrumental in biomonitoring programs for assessing the ecological quality of aquatic ecosystems (Pinto *et al.,* 2021). These alterations, which are reflective of anthropogenic pollutant effects, offer insights into the overall health of the population within the ecosystem. Despite being labor-intensive, histological methods allow individual observation of tissue alterations, establishing direct links with physiological functions, such as growth, reproduction, and respiration.

In marine biomonitoring, histopathological parameters, particularly in the gills and liver, serve as key indicators of lesions induced by chemical contaminants (Pinto *et al.,* 2021). Through detailed examination, histopathological assessments offer crucial insights into the health of aquatic ecosystems, making them indispensable tools for environmental quality evaluations (Yancheva *et al.,* 2016; Pinto *et al.,* 2021).

2.3 Histological alterations in fish gills

Fish gills, constituting over 50% of their surface area, are vital organs involved in multiple functions such as ion transport, gas exchange, and waste excretion. Gills serve as the primary route for waterborne toxicants to enter fish organisms, making them sensitive indicators of the effects of environmental pollutants (Rosseland *et al.,* 2000). Gill metabolism is significantly affected by toxicant incorporation, leading to potential damage, especially as gills accumulate bioavailable pollutants (Yancheva *et al.,* 2016). In teleost fish, gill studies are essential for bioaccumulation research, as they reflect metal speciation and bioavailability in water. Gill surfaces are sensitive indicators of environmental changes, including temperature shifts, acidification due to acid rain, salts, and heavy metals, and alterations in the composition of aquatic environments. Additionally, gill surfaces act as metal-binding ligands, facilitating metal bioaccumulation in water (Drishya *et al.,* 2016).

Gill histopathology serves as a valuable biomarker of environmental contamination in fish. Detailed descriptions of normal gill morphology are available. Gill alterations, which are considered non-specific biomarkers, indicate exposure to diverse organic and inorganic contaminants (Drishya *et al.,* 2016). Edema, lamellar epithelium detachment, and lamellar fusion are common pathologies observed in fish gills following exposure to xenobiotics. Cell proliferation and thickening of filamentous epithelium are adaptive responses that potentially increase mucous secretion and affect respiration and osmoregulation. Rupture of the branchial epithelium and lamellar axis vasodilatation, leading to aneurysms, is a significant response to pollutants, such as heavy metals (Drishya *et al.,* 2016; Pinto *et al.,* 2021). Based on various studies, a broad list of histological changes in ecotoxicological studies is illustrated in Fig. 1. Although gill histopathology is sensitive, meticulous preparation is required.

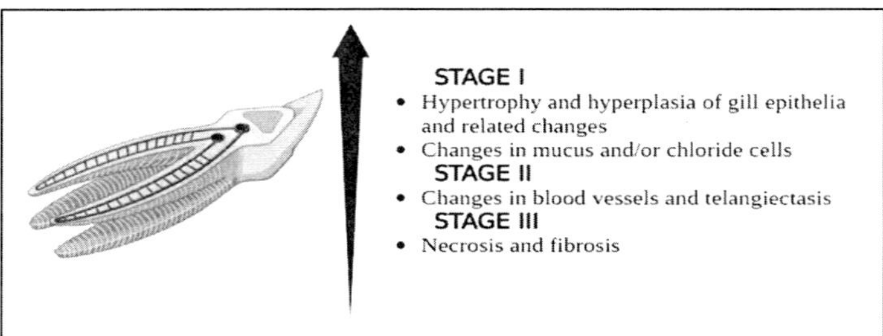

Fig. 1. Gill histological anomalies documented in fish ecotoxicological studies are divided into three stages. Stage I: Most prevalent concomitant with lesser threat to fish, Stage II: Moderately prevalent with moderate threat to fish, Stage III: Least prevalent with greatest threat to fish

In summary, gill histopathology, despite species variation in sensitivity, offers a promising and specific biomarker for environmental contamination in fish. The observed alterations, ranging from edema to epithelial rupture, indicated exposure to various pollutants. However, the time-consuming nature of this technique necessitates careful consideration in research.

2.4 Histological changes in fish liver

Fish internal organs, particularly the liver, are a common target for studying contaminant accumulation and its effects on fish health. Among these, the liver stands out because of its role as a detoxification organ and its large blood supply, which makes it highly susceptible to toxicant exposure (Rosseland *et al.,* 2000). The livers of healthy fish exhibit a typical parenchymatous structure composed of hepatocytes arranged around blood sinusoids. However, exposure to pollutants can lead to various histopathological alterations in the liver, such as hepatocellular vacuolation, necrosis, inflammatory responses, and changes in hepatocyte nucleus size and shape (Pinto *et al.,* 2021). These histopathological changes observed in fish liver tissues exposed to diverse contaminants, including heavy metals, organic pollutants, and pesticides, indicate cellular stress and the potential impairment of vital functions. Hepatic alterations, such as vacuolations, necrosis, and changes in nuclear morphology, often signify increased metabolic activity or pathological origins (Pinto *et al.,*

2021). Liver hyperemia observed in contaminated environments can lead to necrosis and atrophy, indicating disturbances in hepatic blood circulation and potential regressive changes. A detailed list of the documented histological changes in fish liver toxicological studies is shown in Fig. 2. Overall, these histopathological alterations in the liver serve as valuable indicators of pollution effects, reflecting the biochemical disturbances and challenges faced by fish in contaminated environments.

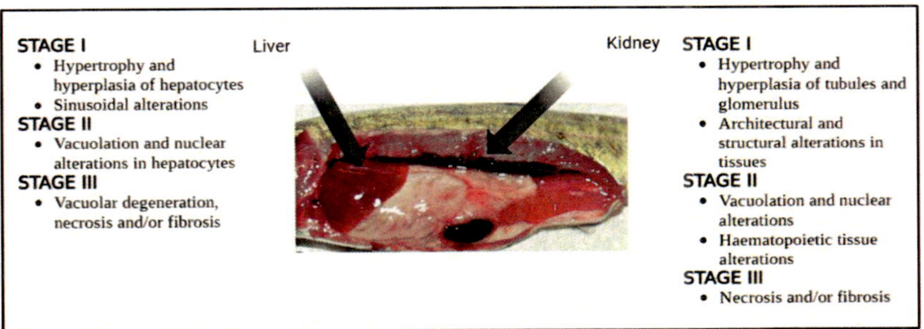

Fig. 2. Broadly documented liver and kidney histological anomalies in fish ecotoxicological studies can be divided into three stages. Stage I: Most prevalent concomitant with lesser threat to fish, Stage II: Moderately prevalent with moderate threat to fish, Stage III: Least prevalent with greatest threat to fish (Image source: Kumar *et al.*, 2018)

2.5 Histological changes in fish kidney

The kidney plays a crucial role in maintaining the body's fluid balance and eliminating waste products, making it a vital organ for fish physiology. Fish kidneys, particularly teleosts, have evolved diverse structures to adapt to different environmental demands. They work alongside the gills and intestines to excrete and maintain an internal balance. In fish, the kidney acts as a route for excreting metabolites and xenobiotics, substances to which the fish may be exposed, and handles nitrogen-containing waste products, such as ammonia and creatinine (Yancheva *et al.,* 2016).

Histological studies of fish kidneys have provided important insights into the effects of environmental pollutants. Various alterations in fish kidneys, such as tubule degeneration, glomerular changes, and dilation of renal tubules, have been observed after exposure to contaminants such as pesticides, heavy metals, and organic pollutants (Yancheva *et al.,* 2016). These changes often reflect the sensitivity of fish to specific substances and severity of environmental pollution. Histopathological examination of fish kidneys, although not as extensively studied as the gills and liver, provides valuable information about water contamination and its effects on aquatic life. A categorized yet precise list of such documented histological anomalies is illustrated in Fig. 2.

By analyzing these kidney alterations, scientists can assess the health of fish populations and gain insights into their overall environmental quality.

Conclusion

In recent decades, histopathological changes observed in key fish organs, such as the gills, liver, and kidneys, have proven invaluable in ecotoxicological research and risk assessment. Although these biomarkers lack specificity, they offer precise insights into the impacts of diverse contaminants, often indicating alterations at lower biological levels. Despite being time consuming and requiring skilled experts, these assessments are relatively affordable. Given their accuracy and ability to detect subtle biological changes, we advocate the increased incorporation of histopathology into monitoring programs for contaminated aquatic environments. When combined with other biomarkers and chemical analyses of water and sediments, histopathological evaluations provide a comprehensive understanding of the ecological health of aquatic ecosystems, enhancing our ability to manage and mitigate environmental pollution effectively.

References

B.O. Rosseland, J.C. Massabuau, J. Grimalt, S. Rognerud, R. Hofer, R. Lackner, I., Vives, M., Ventura, E., and Stuchlik, R. Harriman, and P. Collen, "Fish ecotoxicology," EMERGE Final Report, pp.41-50, 2000.

I. Pinto, S. Rodrigues, O.M. Lage, and S.C. Antunes, "Assessment of water quality in Aguieira reservoir: Ecotoxicological tools in addition to the Water Framework Directive," Ecotoxicology and Environmental Safety, vol. 208, pp. 111583, 2021. doi: https://doi.org/10.1016/j.ecoenv.2020.111583

M. Ferreira, M. Caetano, P. Antunes, J. Costa, O. Gil, N. Bandarra, P. Pousão-Ferreira, C. Vale, and M.A. Reis-Henriques, "Assessment of contaminants and biomarkers of exposure in wild and farmed seabass," Ecotoxicology and Environmental Safety, vol. 73(4), pp. 579-588, 2010. doi: https://doi.org/10.1016/j.ecoenv.2010.01.019

M.K. Drishya, B. Kumari, K.M. Mohan, A.P. Ambikadevi, and B. Aswin, "Histopathological changes in the gills of fresh water fish, Catla catla exposed to electroplating effluent," International Journal of Fisheries and Aquatic Studies, vol. 4(5), pp. 13-16, 2016.

S. Ullah, and M.J. Zorriehzahra, "Ecotoxicology: a review of pesticides induced toxicity in fish," Advances in Animal and Veterinary Sciences, vol. 3(1), pp. 40-57, 2015. doi: https://doi.org/10.1016/S1382-6689(02)00021-2

V. Poleksic, M. Lenhardt, I. Jaric, D. Djordjevic, Z. Gacic, G. Cvijanovic, and B. Raskovic, "Liver, gills, and skin histopathology and heavy metal content of the Danube sterlet (Acipenser ruthenus Linnaeus, 1758)," Environmental Toxicology and Chemistry: An International Journal, vol. 29(3), pp. 515-521, 2010. doi: https://doi.org/10.1002/etc.82

V. Yancheva, I. Velcheva, S. Stoyanova, and E. Georgieva, "Fish in Ecotoxicological Studies," Ecologia balkanica, vol. 7(1), 2015.

Colour Plates

Chapter 2: Modern Biotechnological Approaches of Fish Based Biomaterials for Effective Medicinal Uses: An Update

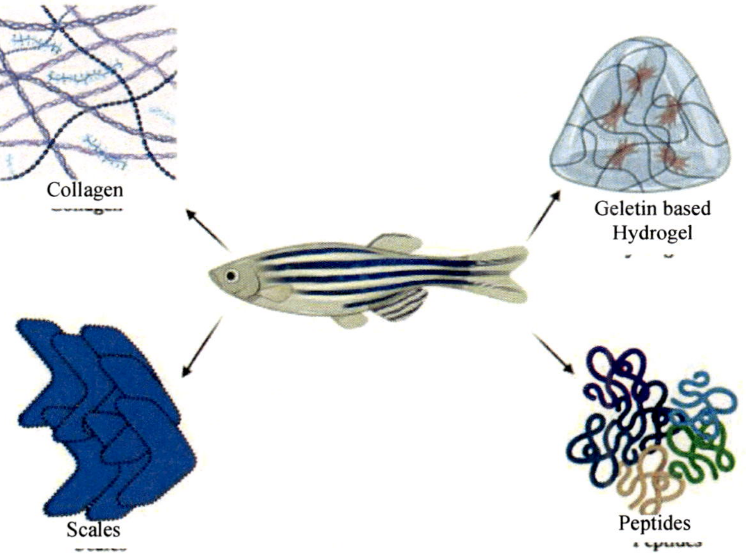

Fig. 1. Diverse materials of marine fish

Chapter 6: Application of Medicinal Plant in Fish Feed and its Challenges

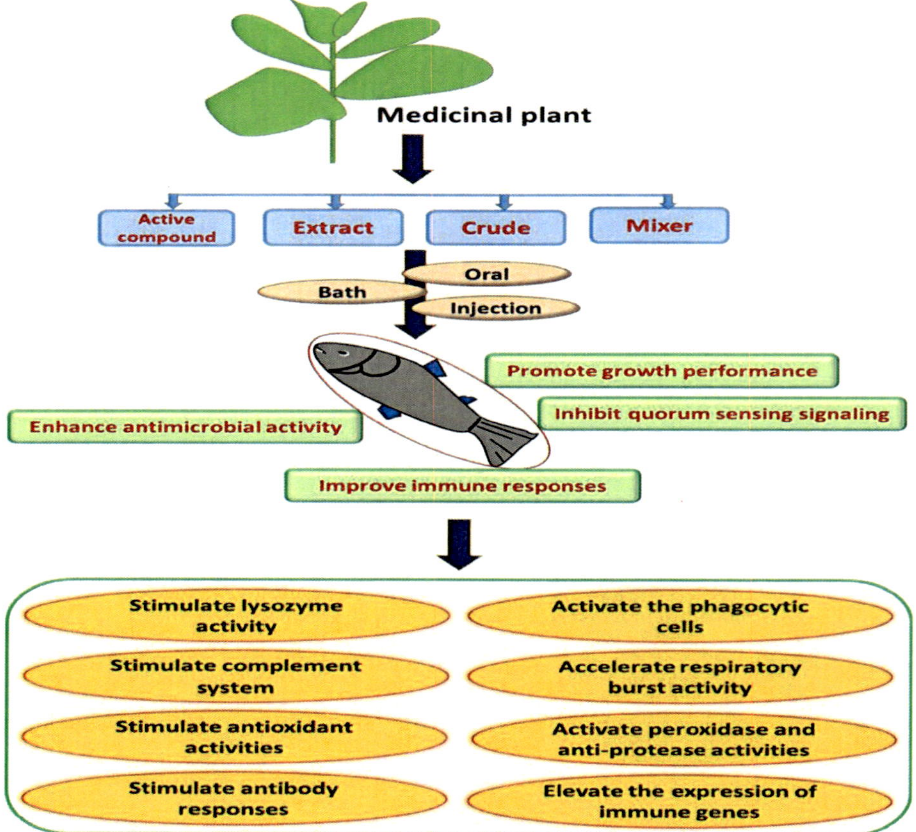

Fig. 1. Role of medicinal plants on immune status in fish

Chapter 8: Fish Therapeutics and Its Limitations

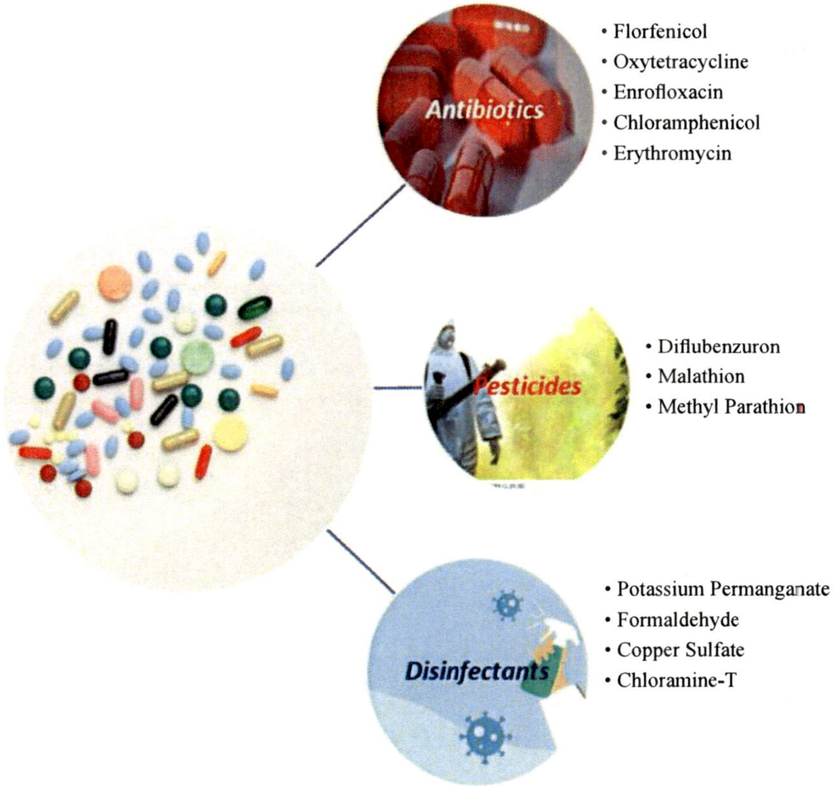

Antibiotics
• Florfenicol
• Oxytetracycline
• Enrofloxacin
• Chloramphenicol
• Erythromycin
Pesticides
• Diflubenzuron
• Malathion
• Methyl Parathion
Disinfectants
• Potassium Permanganate
• Formaldehyde
• Copper Sulfate
• Chloramine-T

Fig. 1. Commonly used fish antioxidants in aquaculture.

Chapter 16: Histological Alterations in Gill, Kidney and Liver of Fish as a Diagnostic Tool for Ecotoxicological Research

STAGE I
• Hypertrophy and hyperplasia of gill epithelia and related changes
• Changes in mucus and/or chloride cells
STAGE II
• Changes in blood vessels and telangiectasis
STAGE III
• Necrosis and fibrosis

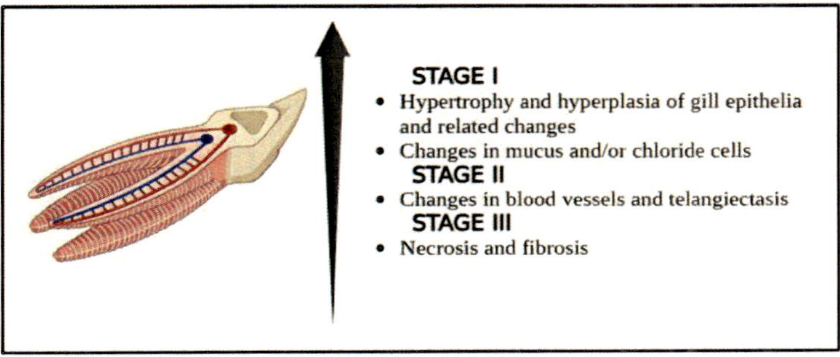

Fig. 1. Gill histological anomalies documented in fish ecotoxicological studies are divided into three stages. Stage I: Most prevalent concomitant with lesser threat to fish, Stage II: Moderately prevalent with moderate threat to fish, Stage III: Least prevalent with greatest threat to fish.

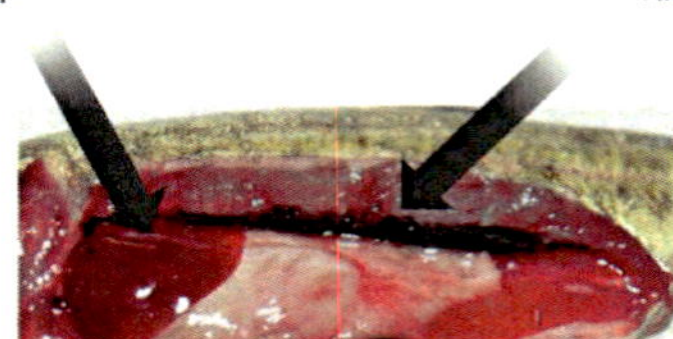

Fig. 2. Broadly documented liver and kidney histological anomalies in fish ecotoxicological studies can be divided into three stages. Stage I: Most prevalent concomitant with lesser threat to fish, Stage II: Moderately prevalent with moderate threat to fish, Stage III: Least prevalent with greatest threat to fish (Image source: Kumar *et al.*, 2018).